The Battle for the Sinister Universe:

The Heuristics

Greg Feild

October 29, 2018

Abstract:

This book is a compilation of the the following four papers:

On math, physics, and metaphysics

On quantum mechanics

On epistemology and ontology !

Toward a metaphysics of mass and motion

On Math, Physics, and Metaphysics:

A Polemic

Greg Feild

October 1, 2018

About the author:

    Greg Feild ~~mourns~~ has hope for humanity!

        He enjoys his privacy.

There is something in the human psyche which, faced with an unbelievable proposition, rushes forward to embrace it, to say 'yes, it *must* be so!', and to rejoice in the ruin of common sense that follows.

. . .

Paradox in our tradition sets the world against itself: the world eats itself up before our eyes; and that is the point of paradox. It seems to show the victory of thought over reality; 'I can believe anything,' it says, 'even this. Join me!'

                                       -- Roger Scruton
                                       Modern Philosophy

(All quotes used without permission … : /)   We claim 'fair use'   :)

Abstract:

This book is a rant and a tirade against modern society and modern physics.

It is not for the feigner, feinter, nor the faint of heart.

Forewarned is forearmed !

What is called thinking?:

This situation [that we are still not thinking] is grounded in the fact that science itself does not think, and cannot think -- which is its good fortune, here meaning the assurance of its own appointed course. Science does not think. This is a shocking statement. Let the statement be shocking, even though we immediately add the supplementary statement that nonetheless science always and in its own fashion has to do with thinking. That fashion, however, is genuine and consequently fruitful only after the gulf has become visible that lies between thinking and the sciences, lies there un-bridgeably.
There is no bridge here --- only the leap.

Hence there is nothing but mischief in all the makeshift ties and asses' bridges by which men today would set up a comfortable commerce between thinking and the sciences.

↖ ↗ ↘ ↙ ↖ ↗ ↘ ↙ ↖ ↗ ↘ ↙

    In the universities especially, the danger is still very great that we misunderstand what we hear of thinking, particularly if the immediate subject of the discussion is scientific. Is there any place compelling us more forcibly to rack our brains than the research and training institutions pursuing scientific labors? . . . if a distinction is made between thinking and the sciences, and the two are contrasted, that is immediately considered a disparagement of science. There is the fear even that thinking might open up hostilities against the sciences, and becloud the seriousness and bespoil the joy of scientific work.

-- Martin Heidegger

Chapter III: Spinors in Three-Dimensional Space
I - The Concept of a Spinor: 52. Definition

Suppose the three-dimensional space E3 is referred to a system of orthogonal coordinates: let (x_1, x_2, x_3) be an isotropic vector, i.e., have zero length. We can associate with this vector the components which satisfy

$$x_1^2 + x_2^2 + x_3^2 = 0, \tag{i}$$

two numbers ξ_0, ξ_1 given by

$$x_1 = \xi_0^2 - \xi_1^2 \tag{ii}$$

$$x_2 = i(\xi_0^2 + \xi_1^2) \tag{iii}$$

and

$$x_3 = -2\xi_0\xi_1 \tag{iv}$$

These equations have two solutions given, for example, by the formula

$$\xi_0 = \pm\sqrt{(x1 - ix2)/2} \quad \text{and} \quad \xi_1 = \pm\sqrt{(-x1 - ix2)/2} \tag{v}$$

It is not possible to give a consistent choice of sign which will hold for all isotropic vectors in such a manner that the solution varies continuously with the vector. Thus, suppose there is such a choice; start with a definite isotropic vector and suppose it to be continuously rotated round $0x_3$ through an angle α: $x_1 - ix_2$ will be multiplied by $e^{-i\alpha}$, thus by continuity ξ_0 will be multiplied by $e^{-i\alpha/2}$. When the angle of rotation is 2π, the isotropic vector returns to its original position, but ξ_0 is multiplied by $e^{-i\pi} = -1$; i.e, its value is of the opposite sign of that originally selected.

The pair of quantities ξ_0, ξ_1 constitute a *spinor*. A spinor is thus a sort of "directed" or "polarised" isotropic vector; a rotation about an axis through an angle of 2π changes the polarisation of this isotropic vector.

:)

-- Elie Cartan
Leçons sur la théorie des spineurs

Chapter 8: Spin. 8.3 **Spinor Geometry** [abridged]

$\langle S \rangle = \hbar/2 \, \chi^\dagger \sigma \chi$

$\chi = (C_+, C_-)$

$C_+ = e^{i\delta} \cos\theta/2 \, e^{-i\varphi/2}$, and $C_- = e^{i\delta} \sin\theta/2 \, e^{+i\varphi/2}$ (vi)

The occurrence of half angles is an important feature of (vi).

...
... start with some arbitrary $\langle S \rangle$, and consider various possible rotations through 2π.
If θ is replaced by $\theta + 2\pi$, since

$\cos(\theta + 2\pi)/2 = -\cos\theta/2, \quad \sin(\theta + 2\pi)/2 = -\sin\theta/2$ (vii)

both C+ and C- change sign. Also, if φ is replaced by $\varphi + 2\pi$, since

$e^{i(\varphi + 2\pi)/2} = e^{-i\varphi/2}$ (viii)

C+ and C- change sign. Under such rotations any spinor changes sign, and so spinors are not single valued.

-- Rolf G. Winter
"Quantum Physics"

409.05.

$\sqrt{-1} = e^{(2r + 1)\pi i/2}$

This square root has two different values, dependending on whether r is even or odd; they are, respectively,

$\cos \pi/2 + i^*\sin \pi/2 = i, \quad \cos 3\pi/2 + i^*\sin 3\pi/2 = -i$ (iix)

409.06.

$\sqrt[3]{1} = e^{2k\pi i/3}$ (ix)

This has three different values:

$e^{2r\pi i} = \cos 0 + i^*\sin 0 = 1$ (x)   [spin]

$e^{(2r\pi + 2\pi/3)i} = \cos 2\pi/3 + i^*\sin 2\pi/3 = -½ + i\sqrt{3}/2 = \omega$ [positive polarization]

$e^{(2r\pi + 4\pi/3)i} = \cos 4\pi/3 + i^*\sin 4\pi/3 = -½ - i\sqrt{3}/2 = \omega^2$ [negative polarization]
   [positive energy]

-- Herbert Bristol Dwight
"Tables of Integrals and Other Mathematical Data"

The universal model of spinor (i.e. lepton) propagation is shown in Figure I (15).

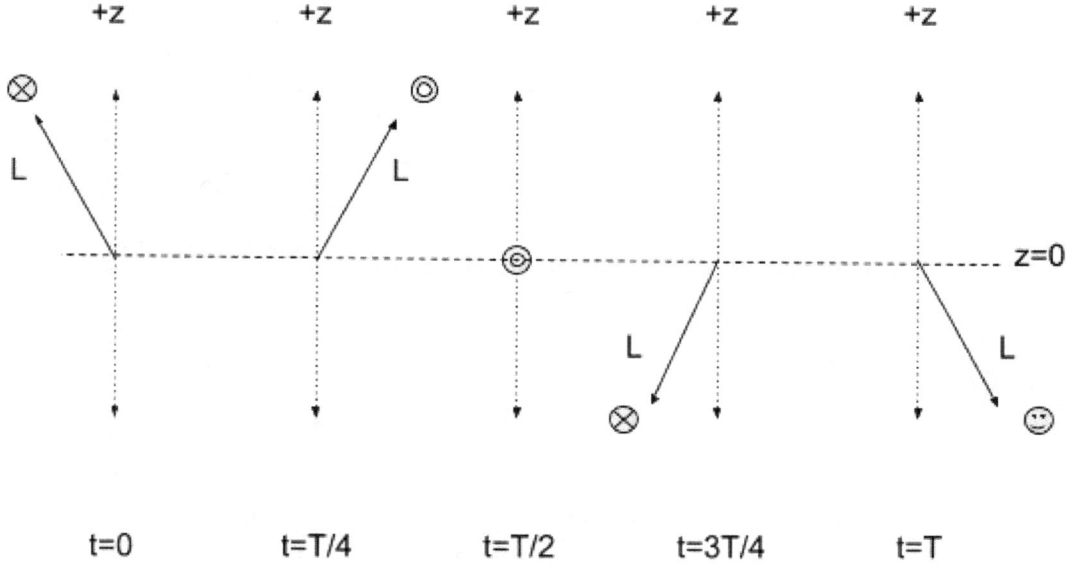

Figure I: At rest with an spinor traveling the the z-direction. The angular momentum vector 'precesses' about the direction of motion, tracing out a closed, three dimensional figure eight. The x symbol represents motion into the page. The spin of the particle completes one 'rotation' after time, t = T/2. A spinor is a zero length vector. Is our 3D figure eight an isotropic cone? What's an isotropic cone? No one knows!   Our cone is polymorphic . . .

The universal 'reverse pilot wave theory' is shown in Figure II (17).

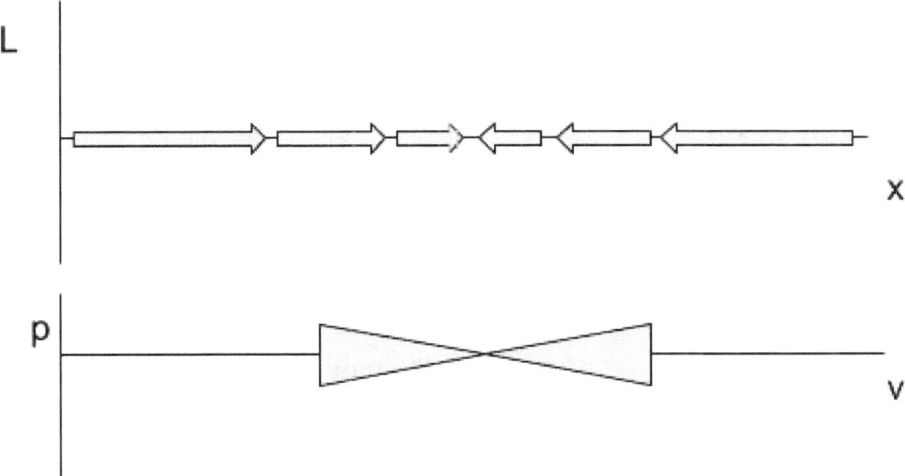

Figure II: Reverse pilot wave theory. Note the isotropic cone !

Spinors:

In our model of lepton, or spinor, propagation, both the particle spin and the angular momentum rotate about the axis of the direction of travel. The spin completes one rotation every 2π radians. The angular momentum vector returns to its original value after 4π radians as illustrated in Figure I. The polarization of a particle changes direction every 2π radians while the particle maintains a constant spin *and* helicity (15,16,17).

In the universal model, particles follow a well defined path while the angular momentum available for interaction varies sinusoidally as dictated by the particle wave function; a (not quite so) simple harmonic oscillator. We call this idea reverse pilot wave theory!

The reverse pilot wave model is depicted schematically in Figure II.

As a particle follows a well defined path, the angular momentum available for transfer during interaction varies, and is projected sinusoidally along the direction of travel. The same is true for the linear momentum of the particle.

For a free particle, $x_0 = 0$, $v_0 = v$

$$x = v_0 t \sin(px - Et)/\hbar \qquad \text{(xi)}$$

$$p = mv_0 \cos(px - Et)/\hbar \qquad \text{(xii)}$$

The variables x and p are out of phase by $\pi/2$.

When the particle polarization is in 'full bloom' as in Figure I at t=0, the particle is best able to scatter (absorb and emit) a real photon, allowing a measurement of the position. There is no ability to measure the momentum at this point.

At time t=T/2, the particle angular momentum is essentially zero. This is the time the particle is most easily reflected from a surface (as it is the time when the particle is most responsive to 'virtual photon interactions') and thus we can measure the particle momentum.

Like all variable pairs subject to the uncertainty relation, the ratio p/x can be predicted and measured exactly

$$p/x \sim \text{ctn}(px - Et)/\hbar$$

We can define the operator p/x in the usual way

$$P/X = -i\hbar/x \, \partial/\partial x$$

↖ ↗ ↘ ↙ ↖ ↗ ↘ ↙ ↖ ↗ ↘ ↙

I spotted the formula for the cubed root of one while thumbing through my table of integrals with the vague hope I might actually integrate something!

The only thing we want to say on this matter at the moment is: intriguing . . .

But ...

It resembles a spinor; and looks like the spin and spin angular momentum of a particle (the only thing missing is a factor of $\hbar$!).

:/

Introduction:

What is wrong with physics today?

The wrong people, do the wrong 'physics', for the wrong reasons.

These people enter the field, already fully convinced of, and enchanted with, the stupid, supernatural explanations, and the supposedly ineffable and inexplicable aspects of modern physics.

This is even *before* they read, study, or 'understand' the incontestable, and incontrovertible, mathematical proofs of modern physics.

Perhaps, we should implement a psychological screening process for entry into this new, oxymoronic field of the professional, theoretical physicist.

Or, perhaps we should no longer pay people, who claim to be clever, to pick their butts.

I am on the fence. Currently, there is no way to ascertain whether anyone is insightful, intelligent, disinterested, and disciplined enough to be paid a *salary*, just to ponder the ineffable, and inexplicable, and somehow come up with The Answer.

Our field will mature one day.

I am convinced.

In the meantime, theoretical physicists should stop calling a press conference every time they have a fart crosswise.

Science <u>journalists</u> need to realize that they must start questioning and critiquing their subjects' curious claims. Currently, they gush about the book, and **fawn all over the author.**

Now is time for Investigative Science Journalism.

Currently, the foxes are in charge of the henhouse.

The rest of us eat cake.

Metaphors are tricky!

Forces and potentials:

In the universal model, the static gravitational force between two particles (e.g. electrons) is the same as the classical, Newtonian force

$$F_G = Gm^2/R^2 \qquad (1)$$

The formula for the static electric, or Coulomb, force has been modified to include 'relativistic' effects

$$F_E = (e/m_e)^2 (1/4\pi\varepsilon)\, m^2/R^2 \qquad (2)$$

The ratio F_E/F_G yields the relative strengths of these two forces in our model

$$F_E/F_G = (e^2/4\pi\varepsilon)(1/Gm_e^2) = \alpha/\alpha_G \qquad (3)$$

where α and α_G are defined in the 'usual way'. Please see references (6,17) for details.

The total force between two electrons (or planets, etc.) in planar coordinates is

$$\mathbf{F}/E_{TOT} = K*(c/R)^2 \mu - K*(\mu v^2/R^2) - K*(l^2/\mu R^3) \qquad (4)$$

where

$$K == (G/c^2 - (e/m_e)^2 (\mu/4\pi)) \qquad (5)$$

In equations (4) and (5) the universal model combines gravity and electromagnetism in a manner parallel with, and *exactly* analogous to, the way Maxwell combined electricity and magnetism; right down to the diminishing factors of $1/c^2$ with each successive term!

↖ ↗ ↘ ↙ ↖ ↗ ↘ ↙ ↖ ↗ ↘ ↙

Imagine an electron and a proton separated by some arbitrary distance. The interaction between the two is completely described by equation (4), from the time they begin moving toward one another until they 'bind' and form a hydrogen atom.

At what point does the electron become 'cloudy'? When does it deviate from rectilinear, planar motion and begin bobbling around randomly, in the absence of any new forces or torques, in three dimensions? When does the proton potential become zero and the electron potential become -17 eV?

We are going to cautiously say never!

The hydrogen atom is planar, just like planetary motion. Physics is physics and space is space and mass is mass. Great or small, everything behaves the same way.

Fascination with size and scale, a human foible, has no place in physics.

↖ ↗ ↘ ↙ ↖ ↗ ↘ ↙ ↖ ↗ ↘ ↙

Consider two identical particles under interaction with an initial separation r_0, and zero initial velocities. $|v_1 - v_2| = |v - (-v)| = 2v$. The force between the two is

$$\mathbf{F} = K\, m^2/(r_0 + 2vt)^2 \qquad (6)$$

$$W = \int \mathbf{F} \cdot \mathbf{dr} \qquad (7)$$

$$W = \int \mathbf{F} \cdot dt\, dr/dt \qquad (8)$$

$$W = \int \mathbf{F} \cdot \mathbf{v}\, dt \qquad (9)$$

$$W = K\int \mathbf{v} \cdot m^2/(r_0 + 2vt)^2\, dt \qquad (10)$$

We remind the reader that equation (10) is the universal Lagrangian.

This is the function to minimize. Note, again, and 'as usual', and as necessary in the universal model, the integral is over the time *only*.

Time is fundamental.

All interaction happens in time and over time.

t,h,c

As a prelude to the next section, let's consider the interaction between our two identical particles (now, electrons) in terms of their potential fields and Green's theorem.

$$\psi(r,t) = kq_1/|r_1-r_2| \; ; \; \phi(r,t) = kq_2/|r_2-r_1| \; ; \; q = m(e/m_0) \quad (11)$$

$$\partial(\psi\phi)/\partial t = 0 = \mathbf{Del} \cdot (\psi \mathbf{Del}\phi - \phi \mathbf{Del}\psi) \quad (12)$$

$$\psi \mathbf{Del}\phi - \phi \mathbf{Del}\psi = 0 \quad (13)$$

$$k^2 q_1 q_2 / |r_1-r_2|^2 - k^2 q_2 q_1 / |r_2-r_1|^2 = 0 \quad (14)$$

which really only expresses the conservation of mass. For gravitational interactions

$$G m_1 m_2 / |r_1-r_2|^2 - G m_2 m_1 / |r_2-r_1|^2 = 0 \quad (15)$$

On current physics:

All phenomena can be reduced to the interaction of mass currents.

Indeed, in our model, the fundamental particle of matter, the lepton, is a closed mass current spinning either to the left or the right. Particles spinning in the same direction repel, and those spinning in opposite directions, attract.

Let's begin with the traditional, non-relativistic, probability current created for a particle with a wave function satisfying the Schrodinger equation

$$\partial \rho / \partial t = i/2m \, (\psi^* .... - .... \psi^*) \quad (16)$$

and create instead, a conservation of the mass current

$$m(\partial \rho / \partial t) = i/2 \, (\psi^* .... - .... \psi^*) \quad (17)$$

We can now write an equation describing the continuity of mass

$$m(\partial \rho / \partial t) + \mathbf{Del} \cdot \mathbf{j} = 0 \quad (18)$$

Where $\mathbf{j}$ is now the mass current and $m(\partial \rho / \partial t)$ is the mass density.

We have resolved the negative probability density problem in a natural way. There is no need to append the electric charge 'by hand' to create a physical current.

The electric charge will enter 'in a natural way' (as well), as the coupling constant, e/m_e, when we include the electromagnetic interaction into our equations.

The fundamental particles and their currents interact in the same way as macroscopic current loops (coils) and linear electric currents (e.g. in wires, etc.). The quantum difference is stricter boundary conditions such as the wave function must be zero at at the boundaries of the system, etc.

As for isospin currents, they too reflect the conservation of mass. In our model, the electric charge is really just one more coupling constant, akin to G and ε.

Consider the interaction between an electron and a neutrino. The total coupling charge, tcc, squared (6), at the electron neutrino vertex is

$$(tcc)^2 = G m_e m_\upsilon - \sqrt{G} k(e/m_0) m_e m_\upsilon \tag{19}$$

$$(tcc)^2 = m_e m_\upsilon (G - \sqrt{G} k(e/m_0)) \tag{20}$$

Electromagnetic induction:

An infinitely long mass current is depicted in a masterful schematic view in Figure 1. If we join the ends together and shrink it to a radius of $r_e = \lambda_e$, the Compton wavelength of the electron, we have an electron. In the limit $r \to 0$, the **E** field goes to zero, and we have the neutrino.

Conversely, the substitution $r = \lambda_\mu$ yields the muon, and the same for the tau.

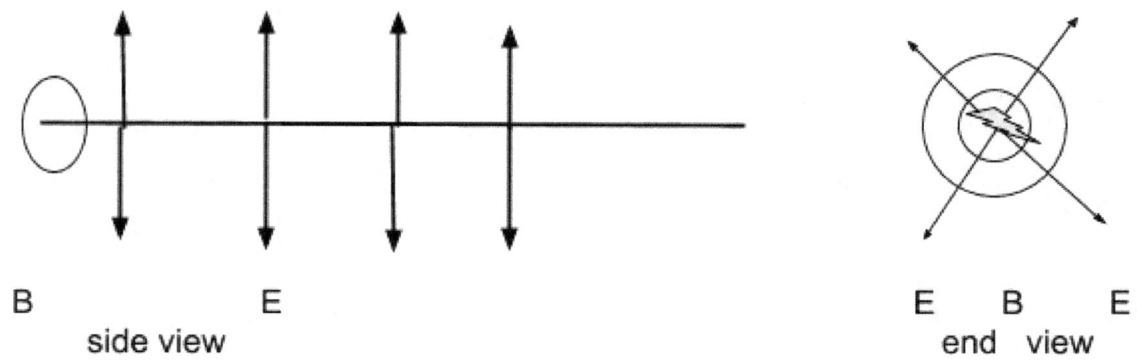

B   E   E   B   E
side view   end view

Figure 1: An 'infinitely' long mass current.

As hinted in our last book (17), we can understand the particle family hierarchy in terms of the quantization of electromagnetic induction. The rest mass energy and rest mass angular momentum of the three charged leptons must satisfy the following relationships

$$\hbar \omega_0 = m_0 c^2 = I_0 \omega_0^2 \qquad (21)$$

$$I_0 \omega_0 = \sqrt{3}/2 \, \hbar \qquad (22)$$

$$I_0 = m_0 \lambda_0^2 / (2\pi)^2 \qquad (23)$$

$$\lambda_0 = h/m_0 c \qquad (24)$$

where $0 = e, \mu, \tau$.

We leave the algebra up to the reader. Send the correct answer to your 'favorite' celebrity physicist!

If electromagnetic induction is quantized, we should be able to measure it.

We don't have any brilliant ideas for such a measurement using electric charge (tiny circuits, perhaps!); or maybe this crazy, new quantum Hall effect we've been hearing so much about.

As for mass currents, perhaps something with neutrino beams, or liquid helium flowing toroidally with 'relativistic' velocity.

I don't know ... I'm from HEP !   :)

The Dirac equation:

The universal Dirac equation is

$$H \psi = \alpha \cdot p \, \psi \tag{25}$$

$$i\hbar \, \partial \psi / \partial t = -i\hbar \, \alpha \cdot \nabla \psi \tag{26}$$

and is satisfied by, or solved with, the universal wave function

$$\psi = \exp(i(p \cdot x - (m-m_0)c^2 t)/\hbar) \tag{27}$$

$$mc^2 \psi = \alpha \cdot p \, \psi + m_e c^2 \psi \tag{28}$$

Obviously, the Klein Gordon equation can be viewed in an equivalent way for the treatment of massive (i.e. 'non-gauge' !) bosons.

Now, we can normalize the Dirac spinors in a physical way. In addition, we needn't fret about, or capriciously toss out, the rest mass term when creating the Dirac (i.e. universal !) Lagrangian.

On math and physics:

Math is a human made invention.

It is a valuable tool, like the torque wrench.
(Who even really understands torque ?)

There is no Platonic realm where numbers, circles, Justice, torque, etc., reside awaiting our discovery and contemplation.

Rather, these objects may exist in a Transcendental realm, where they assist one's reason, as Ideas, in the rational contemplation of concepts.

These machinations are for the satisfaction of the intellect, or Pure Reason, *only*.

No conclusions drawn here may be applied back to the physical world.

That is our reading of Kant, at least.

What is Reason ?

What constitutes 'rational thought' ?

No one knows !

: /

On physics and philosophy:

The areas of philosophy pertinent to the physicist are

    Epistemology: The theory of knowledge,

    Ontology: The theory of the nature of being,

and, if you're a bad physicist,

    Teleology: The study of evidences of design in nature. *

How do we know our theories are real or true? The Copernican theory was chosen over the Ptolemaic because it was more beautiful. Do physicists want to include the fledgling field of Aesthetics to the list? We could also insist the theory explain the data, although this was not a determining factor in the cosmological debate, as each theory was equally accurate.

What is the nature of the objects of physics (e.g. fields) ?

Where do they come from and where do they go ?

Like Batgirl . . .   No one knows!

Some people believe the standard model describes the data, although we would say the data describes the standard model with its 36 knobs.

Some would also say this jerry rigged monstrosity is beautiful.

                                               beauty is in the eye of the beholder

                                                                                :(

*Merriam Webster's Collegiate Dictionary Eleventh Edition

Metaphysics:

In the most ~~boring~~ excellent book ever, "A Critique of Pure Reason", Immanuel Kant demonstrated that the concepts of space, time, and causality, are *a priori*, or innate to the human mind. One hundred million years of evolution can't be wrong!

Kant and his contemporaries believed that the *a priori* concepts of space, time, and causality, plus the 'Principle of Sufficient Reason', were sufficient grounds, *and the only ones we have*, for investigations in natural philosophy.

The idea that physics, operating in space and time, under the assumptions of space, time, and causality, could discover a different, 'truer' explanation of space and time, is absurd *prima facie* (latin is fun!). It is ironic and paradoxical; iradoxical !

Celebrity physicists are not philosophers. They seem to crib their facile and superficial discussions of the Greeks from each other's books. (Just how one writes their thesis!)

The lone exception, in our opinion, is Steven Weinberg. His book "To Explain the World" seemed to be well researched, even "academic"; was interesting, absorbing, and thought provoking. In the beginning of "Gravitation and Cosmology", he casts some doubt on whether the theories of relativity are even "true", then goes ahead and works it all through anyway!

I realize now, we inadvertently cribbed a portion of his essay in the <u>New York Times Review of Books</u> in our book "A critical examination of classical and quantum mechanical waves".

Plagiarism: The sincerest form of flattery !

(Please don't sue!   Dammit Jim!   I'm a doctor not a lawyer !)

:)

Celebrity physicists:

In this section we will only pick on the dead.

Was Stephen Hawking a genius, special, or interesting in any way?
No, he was not.

The man did not understand a lick of physics!

Also, he said many regrettable, completely uneducated, and ignorant things concerning
Issues completely beyond his purview, the bounds of his education and his experience.
What happens now? Disinterment? De-knighting?

Richard Feynman, famous and fabulous, seems the biggest dullard ever to enter the
field of physics. We *curse* him for the worst, most mind boggling, and **stupid** offences of QFT.

Rumor has it, these two people were also unpleasant and odious individuals.
Unfortunately, people will always respect balls over common sense.

Albert Einstein explained Brownian motion and gave us the photoelectric effect.
Then the man went off the deep end; causing irreparable damage to his, our, and
humankind's ability to reason properly and think for ourselves.

People also say he was a bit of a jerk.

Conclusion: Physicists are not special, or any more intelligent, insightful, reflective,
or interesting than the average person.   They are actually kind of horrible.

Tragically, they have no idea. The origin of their misplaced confidence, hubis, giant egos,
and unwarranted sense of superiority is *the only remaining mystery* in this small world of ours.

As for other names: Bell, von Neumann, Everett, Bohm; their books should be
"committed to the flames". We are *not* a fan of book burning but we hate ignorance
so much more.

Anyway, any half wit can earn a degree in physics. *I am* certainly proof of that!

People are free to think what they want, but they should just *keep it to themselves*.

Physicists; please, *please,* **please,** no more freethinking.
Fire your managers, press agents, professional stylists and just go away.

Conclusion:

Reflect before you think.
Think before you speak.
Speak,   as necessary.

On compassion:

We can't stay angry at the celebrity physicist !

You cannot blame a tiger for eating your puppy.   : /

Nor, of course, can one chastise an infant for defecating on itself.

People are messy!

On ethics and morality:

Will the celebrity physicist be compensating all their victims by returning their money?

Can the people who wasted time, money, and brain cells on the celebrity physicist's rot and twaddle bring a class action lawsuit for fraud and deliberate deceit?

Will the celebrity physicist ever feel remorse, embarrassment, or shame?

Probably not in this universe !

:)

Resources:

Quantum Field Theory   (saved our electron magnetic moment bacon !)
Claude Itzykson, Jean-Bernard Zuber

Atomic and Quantum Physics
H. Haken, H.C. Wolf

Modern Elementary Particle Physics
Gordon Kane

Classical Dynamics of Particles and Systems
Jerry B. Marion

Foundations of Electromagnetic Theory
John R. Reitz, Frederick J. Milford, Robert W. Christy

Quantum Physics
Rolf G. Winter

Gauge Theories in Particle Physics
I. J. R. Aitchison and A. J. G. Hey

Quarks and Leptons: An Introductory Course in Modern Particle Physics
Francis Halzen, Alan D. Martin   (most excellent !)

Quantum Field Theory
F. Mandl, G. Shaw

Theoretical Mechanics of Particles and Continua
Alexander L. Fetter, John Dirk Walecka

The Theory of Spinors
Elie Cartan

and

Elementary Modern Physics   (Best Book Ever!) *
Richard T. Weidner, Robert L. Sells

*Except for the omission of c in the formula the the electron magnetic moment!

Books by Greg Feild: The Sinister Universe Series

the pentateuch

1. "A quantum mechanical theory of gravitational interactions"
   CreateSpace Independent Publishing, 8/29/2016

2. "Observations on the quantum mechanical nature of gravity"
   CreateSpace Independent Publishing, 10/8/2016

3. "On gravitation and electric charge"
   CreateSpace Independent Publishing, 10/29/2016

4. "On spin, mass, and charge"
   CreateSpace Independent Publishing, 11/29/2016

5. "On angular momentum, acceleration, and absolute motion"
   CreateSpace Independent Publishing, 1/1/2017

the exegeses

6. "The Sinister Universe"
   CreateSpace Independent Publishing, 3/1/2017

7. "On Parity and Isospin"
   CreateSpace Independent Publishing, 4/11/2017

8. "Reflections on the Sinister Universe"
   CreateSpace Independent Publishing, 5/12/2017

the hermeneutics

9. "On Current Physics"
   CreateSpace Independent Publishing, 6/11/2017

10. "A Critical Examination of Classical and Quantum Mechanical Waves"
    CreateSpace Independent Publishing, 6/18/2017

the gospels :)

11. "On wave particle duality and the quantum of action"
    CreateSpace Independent Publishing, 7/6/2017

12. "On matter, mass, and motion"
    CreateSpace Independent Publishing, 9/14/2017

13. "On action and reaction"
    CreateSpace Independent Publishing, 9/24/2017

14. "A quantum mechanical theory of everything"
    CreateSpace Independent Publishing, 11/5/2017

the expositions

15. "On Interaction"
    CreateSpace Independent Publishing, 4/21/2018

16. "On Rotation"
    CreateSpace Independent Publishing 8/19/2018

the matchbook summary

17. "Revenge of the Sinister Universe: The Reality of Everything'
    CreateSpace Independent Publishing, 9/4/2018

the compilations

   "The Universal Model of Our Sinister Universe: The First Ten Books"
   CreateSpace Independent Publishing, 7/2/2017

   "The Canons of the Sinister Universe:
   The Last Four Books on the Universal Model of Our World"
   CreateSpace Independent Publishing, 11/5/2017

   "The Return of the Sinister Universe: The Immaculate Collection"
   CreateSpace Independent Publishing, 9/4/2018

Last rites:

There is a special place in Hell for the celebrity physicist, along with all the other needy, ignorant, nasty, posers of our world; psychics, pedophile priests, Billy Graham, Mother Teresa, etc., etc., etc., and, 99.9999999999999999999999 % of all politicians.

The holier than thou section is the new, 11th circle of hell.

They had to add an extra section for these malevolent, scum sucking, self serving, attention seeking, unprincipled, *undeserving,* undisciplined, and uneducated egomanics, whose service is only to themselves. They are *literally* destroying society and western culture just to feed their own childish wants and pathetic needs.

Our ignorance is the price.

Knowledge be, <u>damned</u>.

Really, it just makes me want to barf, constantly.

But . . .

Pointing out how much these awful people suck is really a lot of fun!   :)

Schadenfreude

Greg

Rant over.

Let the healing begin !

On Quantum Mechanics

Greg Feild

October 15, 2019

a greg feild theory   :)

↺ ↻
gf

About the author:

I earned a PhD in experimental high energy physics from the Pennsylvania State University working on HERA at DESY in Hamburg, Germany studying photoproduction and deep inelastic scattering in electron-proton collisions.

I did my postdoctoral studies with Yale University working at Fermilab on the CDF experiment at the Tevatron. My primary research interest was particle hadronization in quarkonium production in proton-antiproton collisions.

→- →- →- →- →-→- →- →- →- →-→- →- →- →- →-→- →- →- →- →-→- →- →- →- →-

Despite their good intentions, those very people who believe themselves to be the most faithful spokesmen for their predecessors transform the thoughts which they want simply to repeat; methods are modified because they are applied to new objects. If this movement on the part of philosophy no longer exists, one of two things is true: either the philosophy is dead or it is going through a "crisis." In the first case there is no question of revising, but of razing a rotten building; in the second case the "philosophical crisis" is the particular expression of a social crisis, and its immobility is conditioned by the contradictions which split the society. A so-called "revision" performed by "experts," would be, therefore, only an idealist mystification without real significance. It is the very movement of History, the struggle of men on all planes and on all levels of human activity, which will set free captive thought and permit it to obtain its full development.

Those intellectuals who come after the great flowering and who undertake to set the systems in order or use the new methods to conquer territory not yet fully explored, those who provide practical applications for the theory and employ it as a tool to destroy and to construct --- they should not be called philosophers. They cultivate the domain, they take an inventory, they erect certain structures there, they may even bring about certain internal changes; but they still get their nourishment from the living thought of the great dead. They are borne along by the crowd on the march, and it is the crowd which constitutes their cultural milieu and their future, which determines the field of their investigations, and even of their "creation."

These *relative* men I propose to call "ideologists."

Jean-Paul Sartre
Search for a Method

Abstract:

In this book we examine quantum mechanics in light of
The Universal Model of Our Sinister Universe.

→- →- →- →- →-→- →- –÷- →- →-→- →- →- →- →-→- →- →- →- →-→- →- →- –÷- →-

a thumbnail sketch
a jeweler's stone

                                                                                        a mean dea to call my own

→- →- →- →- →-

standing on the
shoulders of Giants

leaves me cold

a mean idea
to call my own

→- →- →- →- →-

everybody hit the ground
everybody hit the ground
everybody hit the ground
everybody hit the ground

                   -- R.E.M.
                   King of Birds

                                                          Go Dawgs !

Prelude:

In our last book, "On Math, Physics and Metaphysics", we had to admit to a bit of inadvertent plagiarism. I knew I was quoting someone almost verbatim, but I could not remember who it was or where I had absorbed this brilliant piece of prose!

Since we are copping to inadvertent plagiarism, a similar thing happened in the book "On angular momentum, acceleration, and absolute motion" (not the best book ever!).

In the section "Force and acceleration", in the paragraph beginning

"A model where everyone is at rest ...",

the words are almost certainly not mine. While I was typing it in, I felt like I had read it somewhere before, but it sounded like me! I apologize profusely to this cunning linguist. I will probably remember the source one day. In the meantime, perhaps this person may come forward to claim their work.

As far as copyright concerns, and our many book excerpts; we can only argue 'fair use', and their employment in, what we hope is, academic and social advancement.

I understand we've been playing it a little fast and lose here!

But, we are a just a teeny, tiny, little, itty-bitty, Mom and Pop organization.   : (

On the bright side, maybe these folks will get the universal model 'bump'.

                                                                         :)

let's not be litigious !

The phrases "burr in the ass of progress" and "fart crosswise" were coined by the brilliant wordsmith and high school shop teacher Curtis H. Shore, Philipsburg, PA.

Preface:

In our last book, we felt it necessary to say some unkind things, make some unpleasant observations, and resort to coarser language than we'd prefer.   But . .

People either value Knowledge (knowledge for its own sake), or they do not.

People either take "human advancement" seriously, or they do not.

People either define human advancement as the freeing ourselves of superstition, in-group social conditioning, knee-jerk behavior, and tribal inclinations, or they do not.

Some things need to be said.

Unfortunately, in our brave, new, pathetic age of ignorance, people only respond to hysterical screaming and hyperbole.

Why are people constantly shouting and screaming at each other, 24/7, on the television?

That's the way people behave in real life !

:(

turn off the tv. read a book

read a 'real' book

Let the healing resume !

:)

Quantum Physics: **3.7** Operators, Eigenvalues, and Measurements

The state function Ψ contains all the information that can be known about the system. The behavior of Ψ is determined by the Schrodinger equation,

$$H\Psi = i\hbar \partial \Psi / \partial t$$

Each measurable dynaminical quantity is represented by a Hermitian operator Ω with a complete set of orthonormal eigenfunctions Ψ_n

$$\Omega \Psi = \omega_n \Psi_n$$

Since the Ψ_n are complete, the state function can be represented by the expansion

$$\Psi = \sum_n C_n \Psi_n, \quad C_n = \int \Psi_n^* \Psi$$

The possible results of a measurement of Ω are the eigenvalues, and the probability of obtaining ω_n is $|C_n|^2$.

-- Rolf G. Winter

Nothing spooky so far !

The expansion of states above, represents a mathematical superposition of states, and *not* a physical superposition of states. The particle or system is always in a well defined, explicit physical state.

The statistics of quantum mechanics behave exactly like those of any invisible macroscopic physical system, and/or any large collection of correlated variables.

Physicists do not understand statistics ... Who does ?

If you are reading a physics paper and encounter the word 'Bayesian', commit it to the flames!

If you happen across the phrase 'neural network' in a physics book, well,

a fool and their money are soon parted

Introduction:

The world is deterministic. Elementary particle interactions are deterministic.
Physics is deterministic. Quantum mechanics is deterministic.

However, the fact that observer A is now making a measurement at time T,
is not predetermined, although the outcome of their measurement may well be!

So much for determinism.

Laplace's claim to be able to calculate the course of the universe would be undoubtedly true, and assured, were it not for the existence of conscious agents: human beings, dogs, two thirds of the Animal Kingdom, and yes, even the celebrity physicist!

People screw everything up.　　　　　　　They always go *too far*.

When ideas become isms; that is the death of thought.

When abstract ideas become *undiscoverable*, yet certain, concrete realities;
thinking is dead; not only dead . . . but D.O.A.　　　　Is such thought preordained ?

Most of history's most competent philosophers have indicated, or concluded, that it is nigh impossible to truly define thinking (What is it? How do you do it? How do you distinguish thinking from imagining, feeling, or doing syllogisms?), and some would say we are incapable of rational thinking or saying anything meaningful at all!　　(They obviously went "too far" !)

What do *you* think ?

Follow yourself !

The new tao of physics:

The seven stages of grief, and

The twelve step recovery program.

Quantization:

Elementary particles are fundamental units of angular momentum 'oscillating' in two (the photon) or three (the leptons) dimensions, at frequencies determined by the particle energies and momenta; or, more precisely, vice versa !

The fundamental unit of angular momentum is Planck's quantum of action, h.

The oscillations of the quantum of action are manifest as the sinusoidal projection of the particle angular momentum along the direction of travel.

This is the origin, or nature, of the quantization of energy and linear momentum, and the reason for the uncertainty principle.

A particle's position is only available for measurement, or 'real' interaction, for about half the time of any observation period.

The other half of the time, the particle would rather be bouncing off the walls!

These ideas have been explored several times before in previous books, and as we are as loathe to write any more about it (i.e. cut and paste !) as the audience is, most likely, to read it --

Please see "Revenge of the Sinister Universe" and "On Math, Physics, and Metaphysics" for the matchbook, and back of the envelope, summaries.

The only properties of the the elementary particles are spin, parity (!), mass, charge, and helicity. These properties completely define the particle and all particle interactions. These quantities are all measurable and all have physical, if not identical, counterparts in the 'macroscopic' world.

There are no quantum labels. No flavors, no colors, no weak charge, no gluons, and no eternally confined partons or quarks. Partons are electrons and neutrinos (the only new thing under the sun . . .). The neutrino has been hiding inside the neutron the whole time !

Particles cannot be in two places at once, they do not become entangled, and they do not exist in a superposition of states, despite 'measurements' to the contrary.

garbage in, garbage out

(still healing)

Reverse pilot wave theory:

For a free particle, $x_0 = 0$, $v_0 = v$

$x = v_0 t \sin(px - Et)/\hbar$ → $x = A \sin(px - Et)/\hbar$

$p = mv_0 \cos(px - Et)/\hbar$ → $\dot{x} = A(E/\hbar)\cos(px - Et)/\hbar$

The variables x and p are out of phase by $\pi/2$.

In analogy with the case of the classical, one dimensional, simple harmonic oscillator, we can display the variation of the position and the velocity (x, $\dot{x}$) of the particle as a phase space diagram. This phase space diagram will show the relative variations in the ability to measure the position or the momentum of the particle at any time t, and/or the amount of linear momentum available for interaction at any point x, at any time t.

Time, again! Particles move, and systems evolve, in time, and *over* time.
Why doesn't time run backwards?   **Time** doesn't run *any which way*.
Crap in motion, stays in motion.   Anyroad ..

For a free particle, we can write

$x^2/A^2 + (\dot{x})^2/A^2(E/\hbar)^2 = 1$; $A = 1/\sqrt{2}$ (?) ⇒ 1   (why not ?)

$x^2 + (\dot{x})^2/(E/\hbar)^2 = 1$

$d(\dot{x})/dx = -E/\hbar \cdot x/\dot{x}$

For a free particle, the phase space path will consist of closed circles in the x,$\dot{x}$ plane.

For a free particle, "we also know" (17), $T = I\omega^2$

Lagrange's equation is

$\partial T/\partial \theta - d/dt(\partial T/\partial(\dot{\theta})) = 0$; $\dot{\theta} = \omega$; $\omega = E/\hbar$

→ $d/dt(2I\omega) = 0$ → $I \, d\omega/dt = 0$

Which is Newton's second law. More generally

$\tau = dI/dt \, \omega + I \, d\omega/dt$

The hydrogen atom:

Electron orbital decay in the hydrogen occurs via the familiar process of *bremsstrahlung* as illustrated in Figure 1.

Everything is *bremsstrahlung*. Gravitational waves are *bremsstrahlung*.

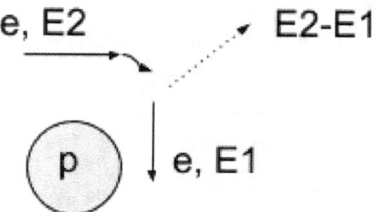

Figure 1: Photon emission in hydrogen electron orbital decay depicted as braking radiation.

An electron in an excited orbital can be considered to be moving tangential to said orbit, and very fast too! The electron is deflected from its path due to the influence of the 'infinitely' heavy proton. (this is the universal complementary principle - sometimes, the proton moves, sometimes, not!)

The emitted photon is represented by the dotted arrow, $E = E_2 - E_1 = h\nu$.

The photon has spin 1, so, of course, for the hydrogen atom as a whole, $\Delta l = 1$.

The electron has shed kinetic energy and is closer to (and as close as it can get!) to its rest mass.

The electron orbits are quantized because the electron *must* form closed elliptical paths.

This is the universal model of electron orbital decay.

The mirror image of our model of beta decay!

Beta decay:

In our model, the proton is a bound state of two positrons and an electron, and the neutron is the bound state of an electron, a positron, and an electron antineutrino.

These assignments would indicate that the proton is 'antimatter' and the neutron 'matter', and thus they would have opposite parity. In our model, the electron, proton, and neutron cannot all have parity = +1 . . .

The strong coupling constant is (6)

$$\alpha_S = \alpha^0_S (1 + \tfrac{1}{2} v^2/c^2 + \tfrac{3}{8} v^4/c^4 + ..) \qquad (1)$$

$$\alpha^0_S = (G/4\pi\varepsilon)^{1/2} (2m_e e/\hbar c) \qquad (2)$$

The weak coupling constant is

$$\alpha_W = (m_v^2 G)/(\hbar c)\, (1 + (v/c)^2 + (v/c)^4 + ...) \qquad (3)$$

Once again (5), we will try to construct a strong force potential using Hooke's law. The force between two charged leptons (i.e. electrons and/or positrons) inside the nucleon is

$$F_{1,2} = \pm\, \alpha_s/R^2 \qquad (4)$$

If we assume R is constant, and neglect the constant term in equation (1), then for an electron and a positron, we have

$$F = \tfrac{1}{2}\, \alpha^0_s v^2/c^2 \qquad (5)$$

$$U = -\, \alpha^0_s v/c^2 \qquad (6)$$

Next, we will want to make the usual expansion of v about the equilibrium velocity v_0, and cue the magic, eventually we can arrive at a covariant formulation in terms of

$$\Delta V^2 = (v - v_0)^2.$$

The proton will be a complicated interaction of two (three ?) strongly coupled harmonic oscillators, which is why we choose to focus on the neutron instead; in particular, we want to investigate the nature of beta decay.

The universal model of beta decay is depicted schematically in Figure 2.

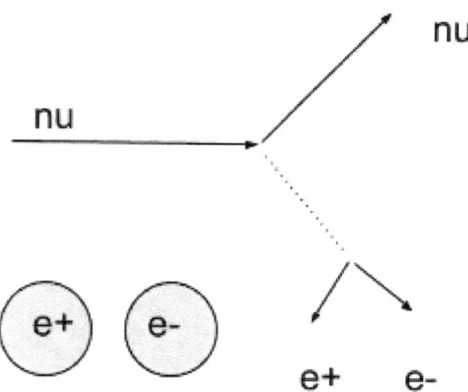

Figure 2: An antineutrino is in a loosely bound orbit around a relativistic positronium nucleus. The nucleus is bound by the potential of equation (6). The nucleus itself is unstable, so the whole thing is a house of cards!

A neutrino, loosely bound 'inside' the neutron, radiates a ~real photon, as if it were free, with the nucleus providing the recoil, of course. The photon decays, or splits, into an electron positron pair. The positron joins with the positronium atom, and we have the proton.

Solid as a rock.   The proton is going nowhere !

It seems we are entering photoproduction territory here. Photoproduction, *bremsstrahlung*, it's all the same thing. Matter shedding energy trying to achieve a state of repose.

(could this be our metaphysics? I enjoy reposing, as well ... so do my dogs !)

As we clearly foresaw over twenty years ago, photoproduction is wave of the future!   :)

Photoproduction experiments and neutrino experiment; it seems they could be done together.

The Dirac equation:

   The universal Dirac equation is

$$H \psi = \alpha \cdot p \, \psi \qquad (7)$$

$$i\hbar \, \partial \psi / \partial t = -i\hbar \, \alpha \cdot \nabla \psi \qquad (8)$$

and is satisfied by, or solved with, the universal wave function

$$\psi = \exp(i(p \cdot x - (m-m_0)c^2 t)/\hbar) \qquad (9)$$

$$mc^2 \psi = \alpha \cdot p \, \psi + m_e c^2 \psi \qquad (10)$$

The universal model is *the* gauge theory anticipated by the standard model !

Rest mass is the global charge and the relativistic mass is the local, or variable charge.

   We salute the constructors of the standard model for keeping this necessary,
   and "beautiful", idea alive.   yay!

   Unfortunately, there were a few misinterpretations on the "meaning" of the mathematics,
and then, of course, everyone took the idea *way too far*, as people *always* do.

   The result: Massive and charged propagators, eight bi-colored gluons, a bunch
of dodgy math ... *et voila* !; you're a television spokesmodel, a prophet, and a futurist !

sweet

   But, back to physics.

   One free particle solution of the Dirac equation for an electron (positive helicity) is

$$\psi = \begin{array}{c} |\,1\,| \\ |\,0\,| \\ \sigma \cdot p/mc^2 \; |\,1\,| \\ |\,0\,| \end{array} \exp(i(p \cdot x - (m-m_0)c^2 t)/\hbar) \qquad (11)$$

   Please forgive the slightly wonky notation. This is the 'spin up' solution.

The solution for the positron (negative helicity) is then

$$\psi = -\sigma \cdot \mathbf{p}/mc^2 \begin{vmatrix} 1 \\ 0 \\ 1 \\ 0 \end{vmatrix} \exp(i(\mathbf{p} \cdot \mathbf{x} - (m-m_0)c^2 t)/\hbar) \qquad (12)$$

also spin up!

There are spin down solutions too, although the helicity of the electron and the positron remain the same.

We no longer need to employ the factor γ^5 to project out left handed and right handed solutions when we are studying leptonic interactions.

Thus, there are no axial vectors in the universal model.

To include the electromagnetic interaction for an electron, we write

$$i\hbar \,\partial \psi / \partial t = -i\hbar \,\alpha \cdot \nabla \psi + (e/m_0 c^2) A^\mu \, i\hbar \,\partial \psi / \partial t \qquad (13)$$

$$i\hbar \,\partial \psi / \partial t \,(1 - A^\mu e/m_0 c^2) = -i\hbar \,\alpha \cdot \nabla \psi \qquad (14)$$

Assume A^μ is due to a second electron

$$A^\mu = -im_2(e/m_0 c^2) \, 1/(v_f - v_i)^2 \, (\phi_f^*(\partial_\mu \phi_i) - (\partial_\mu \phi_f^*)\phi_i) \qquad (15)$$

We have boldly replaced q^2 with ΔV^2

Are we just making junk up as we go along ?

Sometimes !

Close enough for government work !

In equation (15), the 'propagator' term is now in terms of the covariant four velocities of one of the interacting particles. The integral is performed over the variable v_2, from v_i to v_f

This integral should not blow up! There is no need to perform complex integrals to avoid the "mass pole".

The factor A_μ represents the wave function of the virtual photon; $A_\mu \equiv \gamma$.

$$\gamma = \int\int\int\int \exp(ipx - Et)/\hbar \tag{16}$$

where

$$E = m_2 c^2 - m_1 c^2 \tag{17}$$

In this description, there are no fields, no photon "exchange", and the entire interaction is described in terms of the three well behaved wave functions of the three particles involved.

the end

?

On angular acceleration:

    In the grand scheme of things, from the point of view of an universal, inertial reference frame, situated in a galaxy far, far, away, there is no need, or room, for the concepts of linear momentum and rectilinear acceleration. Linear motion is relative.

    Angular acceleration is 'true' acceleration and results in the emission of tangential, retarding radiation, guaranteeing the conservation of energy and momentum.

    In the universal reference frame, the radiating particle and the emitted photon will have equal and opposite angular momenta.

    The world is bathed in photons. From human made electromagnetic radiation, to cosmic rays, to gravitational bremsstrahlung, we are awash in light.

    Light is how we see and communicate!   Light visible and otherwise.

    In physics, we speak of real and virtual photons. There are free photons, bound photons, and binding photons.   We are leptons and radiation.

    We are stardust, billion year old carbon, we are golden, caught in the devil's bargain.
                                                                 -- Joni Mitchell

                                                                                               :)

                         Step into the light !

                 We've got to get ourselves back to the garden !

                             True, dat.

             What's so funny 'bout peace, love, and understanding ? -- Nick Lowe

        nothing   (not really a *non sequitur*)   #introvertsdostuff   !

Conclusion:

People are not rational.   I am not rational !

I have many irrational thoughts a day. Sometimes, six before breakfast !

Let us say rationality consists in identifying, collecting, and ridding ourselves
of irrational thoughts; a *constant* housecleaning, a *victorious* circle of logic,
reason, and reflection.

A task we can only complete together.

A tricky task, as it seems there is no common, human nature to bind and aid us.

Imagine what it must be like to inhabit the fevered, syphilitic, dream of the
celebrity physicist. A place where logic, reason, and good taste go to languish and die.

Impossible !

LOL   :)

Restoration:

Let us return logic, reason, and rationality to the throne; these are the crown jewels of being.
Let common sense, again, guide us in our human, and scientific, endeavors.

Are they not one and the same ? Is there really anything else ?

. . . only succumbing to the mcb

Resources:

Quantum Field Theory   (saved our electron magnetic moment bacon !)
Claude Itzykson, Jean-Bernard Zuber

Atomic and Quantum Physics
H. Haken, H.C. Wolf

Modern Elementary Particle Physics
Gordon Kane

Classical Dynamics of Particles and Systems
Jerry B. Marion

Foundations of Electromagnetic Theory
John R. Reitz, Frederick J. Milford, Robert W. Christy

Quantum Physics (awesome)
Rolf G. Winter

Gauge Theories in Particle Physics
I. J. R. Aitchison and A. J. G. Hey

Quarks and Leptons: An Introductory Course in Modern Particle Physics
Francis Halzen, Alan D. Martin   (most excellent !)

Quantum Field Theory
F. Mandl, G. Shaw

Theoretical Mechanics of Particles and Continua
Alexander L. Fetter, John Dirk Walecka

The Theory of Spinors
Elie Cartan

Elementary Modern Physics   (Best Book Ever!)
Richard T. Weidner, Robert L. Sells

Quantum Mechanics
Claude Cohen-Tannoudji, Bernard Diu, Franck Laloe

Books by Greg Feild: The SInister Universe Series

the pentateuch

1. "A quantum mechanical theory of gravitational interactions"
   CreateSpace Independent Publishing, 8/29/2016

2. "Observations on the quantum mechanical nature of gravity"
   CreateSpace Independent Publishing, 10/8/2016

3. "On gravitation and electric charge"
   CreateSpace Independent Publishing, 10/29/2016

4. "On spin, mass, and charge"
   CreateSpace Independent Publishing, 11/29/2016

5. "On angular momentum, acceleration, and absolute motion"
   CreateSpace Independent Publishing, 1/1/2017

the exegeses

6. "The Sinister Universe"
   CreateSpace Independent Publishing, 3/1/2017

7. "On Parity and Isospin"
   CreateSpace Independent Publishing, 4/11/2017

8. "Reflections on the Sinister Universe"
   CreateSpace Independent Publishing, 5/12/2017

the hermeneutics

9. "On Current Physics"
   CreateSpace Independent Publishing, 6/11/2017

10. "A Critical Examination of Classical and Quantum Mechanical Waves"
    CreateSpace Independent Publishing, 6/18/2017

the gospels   :)

11. "On wave particle duality and the quantum of action"
    CreateSpace Independent Publishing, 7/6/2017

12. "On matter, mass, and motion"
    CreateSpace Independent Publishing, 9/14/2017

13. "On action and reaction"
    CreateSpace Independent Publishing, 9/24/2017

14. "A quantum mechanical theory of everything"
    CreateSpace Independent Publishing, 11/5/2017

the expositions

15. "On Interaction"
    CreateSpace Independent Publishing, 4/21/2018

16. "On Rotation"
    CreateSpace Independent Publishing 8/19/2018

the matchbook summary

17. "Revenge of the Sinister Universe: The Reality of Everything'
    CreateSpace Independent Publishing, 9/4/2018

the diatribe

18. "On Math, Physics, and Metaphysics"
    CreateSpace Independent Publishing, 10/1/2018

the compilations

"The Universal Model of Our Sinister Universe: The First Ten Books"
CreateSpace Independent Publishing, 7/2/2017

"The Canons of the Sinister Universe:
The Last Four Books on the Universal Model of Our World"
CreateSpace Independent Publishing, 11/5/2017

"The Return of the Sinister Universe: The Immaculate Collection"
CreateSpace Independent Publishing, 9/4/2018

Notes:

hell is other people

what can you do ?

try to enjoy !

there is only <u>now</u>

:)

On Epistemology and Ontology

Greg Feild

October 21, 2018

A plea for sanity:

Fellow Kwestors !

Roy G. Biv, here. I have seen it *all*. I've been around since the first rainbow !
Adam named the animals, but *I* named the colors; the concept of colors,
the concept of naming, the concept of having conceptions . . . well . . .
we're still working on most of *these*.   Inventing words is hard !

I invented the words for the colors, using the letters from my name:
red, orange, yellow, green, blue, indigo, and violet.
No easy task when you're not assigned any vowels!

A certain 'Sir' Isaac Newton **stole my thunder** approximately 5000 years later!
Do. Not. Even. Get. Me. Started. . . . although he did realize light must be corpuscular.

Now, most folks don't know this, but I also invented the mnemonic,
My Very Educated Mother Just Served Us Nine Pizzas . . . arrgh!
                                              . . . don't even get me started.
Here is what I have learned over the years:

People think that what they believe is correct.
People believe that *all others* should think as they do.
People want to *impose their will* on *all others*.   Even people they don't know !

This last item is certainly a bad thing; the very *definition* of immorality. As my
daddy used to say, "People have their hands full just taking care of themselves."

People want to be famous. The purpose of the Greek *polis* was to allow citizens to
achieve immortal fame for their oratory skills. David Hume described how people
want to be famous, and why they admire the famous, over 400 years ago.

I don't get it. I personally enjoy the solitude of nature, and not the salacious nature
of other people's business. Our current situation is ***not going to improve***, until
human beings *overcome all four* of the tendencies listed above. Until then, MYOB !

                                            Taste the rainbow,
                                              Be the rainbow,
                                                Roy G. Biv

Abstract:

This is a book about physics *and* philosophy; natural philosophy.

It is mostly about the universal model and the sinister universe !

Inventing book titles is hard.

About the author:

Greg Feild enjoys reading and writing the occaisional book.

Even his dogs think, "he doesn't do anything!"   :)

let s do philosophy !

Epistemology: The theory of knowledge

In this book, we consider the titular philosophical concepts in terms of physics, and in the examination of the concepts of our new model.

In our model, there are three distinct objects: the neutrino, the electron, and photon.

Four, if you include the virtual photon, as we are wont, but wary, to do.

These three particles are **real**, and have physical properties that we can measure.

We **know** they exist.

(We hear the neutrino is becoming 'easier to detect', which is wonderful news.)

The phenomenology of photoproduction will probably show us that the division between real and virtual photons is mostly operational; both practically and mathematically.

We **know** elementary particles exist and we **know** they have five well defined, measurable physical properties: spin, mass, charge, parity, and helicity.

What we **don't know** is whether photons gyre and leptons gimble, (an excellent model!)

and, don't even get us started on the momegraths !

On the absurd:

Go outside. Look to the sky. Find the moon. What is it ? What is it doing there ? Why are we *all* so complacent ? Go out later. The moon has moved ! Crazy.

After my first paper, I contacted a few old friends and told them I had united the four forces. This announcement went over like a lead zeppelin! I was confused since I am not known to lie.

Is it absurd that I devised the theory of everything at my kitchen table and wrote my first five books using Google Docs on a $125 RCA tablet from Walmart?

I don't think so.

Ontology: The theory of the nature of being

Visible radiation allows us to see. Ultraviolet radiation causes sunburn and skin cancer.

Shuffle across the rug in your stockings, touch the doorknob, and you will see and feel a spark! Electricity flows through our electrical wires, through our dendrites, nerves, axons, brains. Electrical currents can tickle or kill a person.

These things are as **real** as the day is long.

What are they 'really' like ?

The electron is a physical object defined by five physical, mathematical quantities: angular momentum, linear momentum, mass, charge, and helicity.

One can completely ***physically*** *describe* a kickball or the earth with the same five quantities. Most people would like to know a little more. Is the kickball red?

There is no more to know about the electron.

Despite claims of mathematical rigor, *and* the adamant claim that all the wacky conclusions of the standard model stem from making sense of the math, the standard model leaves a lot of loose ends dangling.

Examples include: Ignoring the double valued nature of the spinor, not accounting for particle spin (making it a 'quantum label' → flavor, color, weak charge, lepton number, baryon number, strangeness, parity, etc., conservation; a lot of luggage (tags) for a point particle to carry around!), putting things in 'by hand' (e.g. particle charge), the unphysical normalization of the Dirac spinor, the inconsistent treatment of commutators when quantizing leptons, the Pauli exclusion principle, and much, much more !

It seems we have finally answered *the second greatest question of the universe !*

What could "have your cake and eat it too" even possibly *mean* ?

The universal model explains it all!

Metaphysics: The study of the fundamental nature of reality (no theories; systems !)

. . . when using a variational procedure to obtain Lagrange's equations, it is convenient to ignore temporarily the fact that we are dealing with a *physical system* whose motion is completely determined and subject to no variation, and to consider instead only a certain abstract *mathematical* problem. Indeed, this is the spirit in which any variational calculation relating to a physical process must be carried out. In adopting such a viewpoint, we must not be overly concerned with the fact that the variation procedure may be contrary to certain known physical properties of the system. (For example, energy is in general not conserved in passing from the true path to the varied path.) A variational calculation simply tests various *possible* solutions to a problem and prescribes a method for selecting the *correct* solution.

-- Jerry B. Marion
Classical Dynamics of Particles and Systems

Even so long after Maxwell's theory of the (classical) electromagnetic field, the concept of a 'disembodied' field is not an easy one; and we are going to add the complications of quantum mechanics to it. ... At the end, we shall --- like Maxwell --- throw away the 'mechanism' and have simply quantum field theory, the new fundamental reality which has no need of any further underlying framework.   (page 85)

In relativistic quantum mechanics, therefore, we must beware of thinking of the vacuum, too naively, as being 'nothing'. Even when a particle is supposedly free and propagating in the vacuum, the parameters describing it (such as the charge) are changed, from the values appearing in the one particle Hamiltonian, to ones which include the effects of its interactions with the virtual particles of the vacuum. It is appropriate to mention at this point that further remarkable properties of the vacuum seem to be necessary in order to understand fundamental aspects of both the theory of weak interactions (namely, how gauge bosons can acquire mass) and QCD (the problem of confinement).   (page 205)

-- I.J.R. Aitchison, A.J.G. Hey
Gauge Theories in Particle Physics

Introduction:

The ancient pagans believed there was a fundamental difference between the terrestrial and celestial spheres. The stars were not composed of common matter; but rather, stellar stuff.

Then, Isaac Newton united the heavens and the earth.   yay!

What Newton set in place, modern physicists have set asunder!   :/

Outer space is, once again, completely foreign to space here on earth; it curves, it expands, its riddled with dark matter and entwined with dark energies.

really ?   seriously ?   sadly, yes.   :(

boo !

The Schrodinger equation:

The relativistic Schrodinger equation is

$$i\hbar \partial \psi / \partial t = -ic\hbar \nabla \psi \qquad (1)$$

and is satisfied by, or solved with, the universal wave function

$$\psi = \exp(i(\mathbf{p} \cdot \mathbf{x} - (m-m_0)c^2 t)/\hbar) \qquad (2)$$

$$mc^2 \psi = pc\, \psi + m_0 c^2 \psi \qquad (3)$$

We can form the probability current density in the usual way

$$-i\psi(i\hbar \partial \psi^*/\partial t + ic\hbar \nabla \psi^*) + i\psi^*(i\hbar \partial \psi/\partial t + ic\hbar \nabla \psi) = 0$$

$$\Rightarrow -i(\psi^* \nabla \psi - \psi \nabla \psi^*) = \mathbf{j} \qquad (4)$$

Unfortunately, in this formulation, we lose the explicit mass term (from $p^2/2m$) that makes this expression correspond to a physical current (18), and we really don't want to go back to putting the mass and/or charge in "by hand". It's always something!

Our other choice for the relativistic Schrodinger equation is

$$i\hbar \partial \psi/\partial t = -(1/m)(\hbar)^2 \nabla^2 \psi \qquad (5)$$

The mass term is the *relativistic* mass. We insert the wave function

$$mc^2 \psi = p^2/m\, \psi + m_e c^2 \psi \qquad (6)$$

Now, we have a (the?) mass term back. We also have

$$p^2/m = pc \quad \rightarrow \quad p = mc \qquad (7)$$

For a particle at rest, we have $p_0 = m_0 c \equiv \Delta p$, and if we define $\Delta x \equiv \lambda_0$, then

$$\Delta p\, \Delta x = m_0 c\, (h/m_0 c) = h \qquad (8)$$

The exact uncertainty principle! :)

The uncertainty principle:

The uncertainty principle is due to the spinor nature of the electron (i.e. leptons), and the fact that even a particle at rest is constantly flipping polarization, yielding an effective radius or uncertainty in position equal to the Compton wavelength.

A similar argument can made for the photon which is also constantly flipping.

We cannot measure the position and the momentum of a photon, or of a lepton, simultaneously, because they gyre and gimble on the way!

Now, the results of the last section were very exciting, but in the universal model we are more interested in frequency, and changes in frequency, as time determines all things.

Without time, you are nowhere!

From the relationship, $p_0 = m_0 c$, we can see the electron gimbles at the speed of light, so

$$c \Delta t = \lambda_0 \tag{9}$$

and

$$\Delta E = h \Delta \nu = h c / \lambda_0 \tag{10}$$

combine

$$\Delta E \, \Delta t = (h c / \lambda_0)(\lambda_0 / c) = h \tag{11}$$

and we have the other exact uncertainty principle.

The electron gimbles at the speed of light regardless of the linear velocity.

$$p = mc = mc^2 / (c^2 - v^2)^{\frac{1}{2}} \tag{12}$$

Equation (12) expresses the available 'impulse', or the amount of momentum available for transfer by a 'virtual photon', if the electron interacts over any time, t.

This is the "meaning" of the Lorentz invariant quantity, or the spacetime invariant, ΔS^2.

It is the distance 'traversed' by the virtual photon.

The Klein Gordon equation:

The universal wave function of equation (2) satisfies, and *requires*, a 'massless' Klein Gordon equation

$$(-\partial^2 \psi / \partial t^2 + \nabla^2 \psi) \hbar^2 = 0 \quad (13)$$

so, we can use the same equation to describe real photons, spin 1 bosons, and 'spin-less' leptons. Obviously, this equation is not appropriate for 'virtual photons' as the right hand side would not be zero!

For a photon in a gravitational potential, we choose to use the relativistic Schrodinger equation of equation (1) instead ! (We now have two relativistic Schrodinger equations, one for photons, and one for leptons, equation (5))

$$i\hbar \partial \Gamma / \partial t = -ic\hbar \nabla \Gamma + V(x) \Gamma \quad (14)$$

$$E\Gamma = pc\Gamma + V(r) \Gamma \quad (15)$$

with solution

$$\Gamma(x,t) = A \exp(-i(px - (E - V(x))/\hbar) \quad (16)$$

For a free photon, the relativistic Schrodinger equation is

$$i\hbar \partial \Gamma / \partial t + ic\hbar \nabla \Gamma = 0 \quad (17)$$

and we can form the flux and current density equation in the usual way ($c, \hbar$)

$$-i\Gamma(i\partial \Gamma^*/\partial t + i\nabla \Gamma^*) + i\Gamma^*(i\partial \Gamma/\partial t + i\nabla \Gamma) = 0 \quad (18)$$

It seems there is no more need for the KG equation!

:(

Conjugate variables:

This section is totally cribbed from "Classical Dynamics of Particles and Systems (2nd Ed.)"
by Jerry B. Marion; an excellent book, of course !

In our last book, we introduced the idea of plotting the "measurability" of a free particle's position and momentum as a function of time as a phase diagram in the (moving) particle configuration space; x, x^{dot}. Jargony! These variables represent the local projection of the quantities about the point of travel (the center of mass) along the direction of travel.

The more natural choice for the conjugate variables would be: (θ, θ^{dot}) aka (θ, ω).

Since our free particle is traveling with a <u>fixed</u> velocity along the z axis, there are two pairs of conjugate variables: (θ p_θ) ; (z, p_z) ; with one constrained (p_θ).

If we know the initial position and velocity of our particle (and we do!), we can plot its predetermined course, *and the probability for it to interact* at any point x, and at any time t, with a phase diagram in Hamiltonian phase space (page 230, FIG. **7-4**) [particle on a cylinder]

"Thus the phase path on any surface H = const. is a *uniform elliptic spiral, ...*"

More jargon!

The kicker is; if we have a boatload of 'free' elementary particles; quantum mechanics
'reduces' to statistical mechanics.

stat mech !

meh

: /

The Dirac equation:

Mathematically, the electron is represented as a four component column vector multiplied by a time and space dependent wave function.

The photon is represented as a four component column vector multiplied by a time and space dependent wave function.

The four components of a particle's column vector, represent the four conserved scalar quantities of the universal model: the three *independent* (angular) momenta, and the one *common* energy, that *completely characterize a particle*.

Or, spin, mass, charge, and helicity!

The wave function factor, represents the changes in the three momenta and the energy of a particle, as a function of position and time.

On math and physics:

Angular frequency is one of those brilliant ideas of mathematics that may be considered truly Beautiful.

It frees the expression of degrees from the dictates of geometry, and naturally introduces π in 'radians' into our math and physics, as a measure of angle, circularity, and most importantly, for the description of cyclic motion (i.e., 360° vs. $2\pi, 4\pi, 6\pi$, etc.)

Whoever invented the radian was a Genius.

Roger Cotes: Editor of the *Principia*.   Thanks, Wikipedia !

In our model, elementary particles gyre and gimble as they speed through space and time.

A proper, 'unitless', *frame independent*, measure of how far a particle travels in a particular time would then be, the radian, rather than the meter.

On the other hand, everything is geometry! Euclidean geometry:

conic sections → algebra → trigonometry → $e^{i\pi}$ → calculus → conic sections

Rather than the regular polyhedrons; it seems the world is composed of conic sections:

planetary orbits, elementary particle scattering trajectories, elementary particle propagation

Just because we can conceive of non-Euclidean geometries, this does not mean they apply.

parallel lines will never cross !

there, we said it !

: 0

A universal metaphysics ?:

The fundamental unit of matter is the neutrino.
The neutrino is a gravitational-point-mass-charge;
or, a point, gravitational, magnetic moment, μ

$$\mu = h^{bar}/2c \, (1 + \tfrac{1}{2} v^2/c^2 + \ldots) \tag{19}$$

The elemental unit of matter is the dipole! This may spark thoughts of 'dialectic' in our philosophical readers; however, the magnetic dipole is represented, mathematically and schematically, as <u>closed loops</u> of 'magnetic field lines' ! This picture of the neutrino is illustrated in Figure 1.

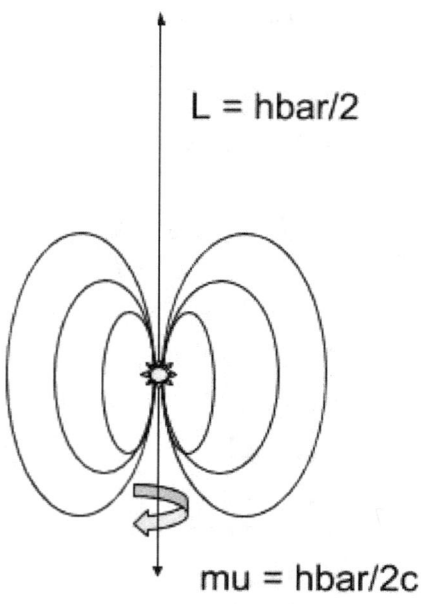

Figure 1: The angular momentum, magnetic moment, and magnetic field lines of a point, gravitational, mass dipole; aka the neutrino!

The neutrino is spinning to the left, with the canonical definitions for the directions of L and μ given by the standard model.

Thus, there are no magnetic monopoles !

Hence; no dichotomy, no disharmony, no dialectical struggle.

:/

In our model, the world is composed of mass currents; open and closed.

These mass currents attract and repel one another.

Why do they do this ? Well, . . . they have to do something !

        otherwise, we wouldn't be here   :)

Whatever the reason, they seem to do it in the simplest, most beautiful, way possible, involving nothing more mathematically, than the 'scalar' and 'cross 'products of vectors.

Only the spinor throws a spanner in the works.

Of course, energy and momentum are conserved, with each particle acquiring equal and opposite momentum, and *'kinetic energy'* during an interaction.

The conservation of energy and momentum now seem no-brainers. Perhaps we should file them with space, time, and causality; as innate to animals, developed for understanding and navigating our environments.

Chirality plays an important role in chemistry and biology.   Is life sinister ?

Let's see: Amino acids are left handed, DNA is right handed, digestible sugars are right handed … We enter all our data into a quantum computer running a Bayesian analysis on neural nets, punch in the desired result, and …..

        Life is Sinister !

Coming soon: "Toward a Metaphysics of Mass and Motion"

Preorder today !

        just joking

Please look for it in 2019, if you are so inclined -- and thank you for your support!

The future of physics:

It is a tired and time worn cliche --- but only because it's true !

:/

Corralling physicists is like herding cats.

They need a strong hand.   They should not be in charge of themselves.

They should not be put in a box.   Neither should cats.

When *everyone* is the smartest person in the room, it is tiresome.   **THONK !**

Here are our recommendations:

Cut the number of physics conferences per year by one jillion.
The physics conference is *the* very definition of a boondoggle.
Only the ski conferences are real.

Computers, computer monitors, scanners, printers, etc.; must last *five years* !

At their one conference a year, the physicists would asses their common priorities, resources, and goals.   Many breakout sessions later, they would produce a *single* document for the NSF, DOE, usw. They would also produce something decipherable for the politicians and the public as well.   Maybe even the same document!

Representatives from the "physics community" would meet with the government
and hash out funding, expectations for the year, a five year plan, a ten year plan, etc.

The participating institutions would then 'bid' on the various projects; committees,
etc., would then allocate the funding and enumerate the responsibilities of all.

Retire the role of the "theoretical physicist". Once employed to *help* 'do the math',
they did nothing but the math, and crap math, at that !   cat not in box, but in hat.

The current crisis;

There are two alternative solutions to our current crisis:

1. Educate everyone.
2. Educate a select few and burn *all* the books.

Group theory:

    People like to put things into groups!

    People have to put things into groups in order to form concepts. Even other people.

On seriousness:

    Seriousness is not the gnashing of one's teeth and the wringing of one's hands;
    this is silliness.

Aliens and Uranus:

    Alien probes are pointless -- they have no elbows!

On leading physicists:

    Who are these leading physicists we keep hearing so much about ?
                                Whom are they leading where ?

---

    cowed by dogs
    suffered by fools
    chuted and schleped
    herd rule   gf

                Everyone's a superhero
                Everyone's a Captain Kirk

                              -- Nena
                              Ninety nine red balloons

Conclusion:

One hurdle to learning and understanding physics is there are a certain number of words and phrases that are necessarily 'technical', esoteric, and arcane; even after we have swept away all the lingo and the jargon.

I personally have read the definition of "Hermitian", like, a billion times, and I still can't remember what it means! If all operators are Hermitian, then, perhaps we needn't mention it all the time!

Sometimes, it seems, the special vernacular is used as window dressing, or to obfusticate;

       If you can't dazzle them with brilliance, . . .

We say:

keep it simple !

physics is fun !!

                                              :)

Resources:

Quantum Field Theory   (saved our electron magnetic moment bacon !)
Claude Itzykson, Jean-Bernard Zuber

Atomic and Quantum Physics
H. Haken, H.C. Wolf

Modern Elementary Particle Physics
Gordon Kane

Classical Dynamics of Particles and Systems
Jerry B. Marion

Foundations of Electromagnetic Theory
John R. Reitz, Frederick J. Milford, Robert W. Christy

Quantum Physics (awesome)
Rolf G. Winter

Gauge Theories in Particle Physics
I. J. R. Aitchison and A. J. G. Hey

Quarks and Leptons: An Introductory Course in Modern Particle Physics
Francis Halzen, Alan D. Martin   (most excellent !)

Quantum Field Theory
F. Mandl, G. Shaw

Theoretical Mechanics of Particles and Continua
Alexander L. Fetter, John Dirk Walecka

The Theory of Spinors
Elie Cartan

Elementary Modern Physics   (Best Book Ever!)
Richard T. Weidner, Robert L. Sells

Quantum Mechanics
Claude Cohen-Tannoudji, Bernard Diu, Franck Laloe

Books by Greg Feild:

1. "A quantum mechanical theory of gravitational interactions"
   CreateSpace Independent Publishing, 8/29/2016

2. "Observations on the quantum mechanical nature of gravity"
   CreateSpace Independent Publishing, 10/8/2016

3. "On gravitation and electric charge"
   CreateSpace Independent Publishing, 10/29/2016

4. "On spin, mass, and charge"
   CreateSpace Independent Publishing, 11/29/2016

5. "On angular momentum, acceleration, and absolute motion"
   CreateSpace Independent Publishing, 1/1/2017

6. "The Sinister Universe"
   CreateSpace Independent Publishing, 3/1/2017

7. "On Parity and Isospin"
   CreateSpace Independent Publishing, 4/11/2017

8. "Reflections on the Sinister Universe"
   CreateSpace Independent Publishing, 5/12/2017

9. "On Current Physics"
   CreateSpace Independent Publishing, 6/11/2017

10. "A Critical Examination of Classical and Quantum Mechanical Waves"
    CreateSpace Independent Publishing, 6/18/2017

11. "On wave particle duality and the quantum of action"
    CreateSpace Independent Publishing, 7/6/2017

12. "On matter, mass, and motion"
    CreateSpace Independent Publishing, 9/14/2017

13. "On action and reaction"
    CreateSpace Independent Publishing, 9/24/2017

14. "A quantum mechanical theory of everything"
    CreateSpace Independent Publishing, 11/5/2017

15. "On Interaction"
    CreateSpace Independent Publishing, 4/21/2018

16. "On Rotation"
    CreateSpace Independent Publishing 8/19/2018

17. "Revenge of the Sinister Universe: The Reality of Everything'
    CreateSpace Independent Publishing, 9/4/2018

18. "On Math, Physics, and Metaphysics"
    CreateSpace Independent Publishing, 10/1/2018

19. "On Quantum Mechanics"
    CreateSpace Independent Publishing, 10/15/2018

compilations

"The Universal Model of Our Sinister Universe: The First Ten Books"
CreateSpace Independent Publishing, 7/2/2017

"The Canons of the Sinister Universe:
The Last Four Books on the Universal Model of Our World"
CreateSpace Independent Publishing, 11/5/2017

"The Return of the Sinister Universe: The Immaculate Collection"
CreateSpace Independent Publishing, 9/4/2018

I earned a PhD in experimental high energy physics from the Pennsylvania State University working on HERA at DESY in Hamburg, Germany studying photoproduction and deep inelastic scattering in electron-proton collisions.

I did my postdoctoral studies with Yale University working at Fermilab on the CDF experiment at the Tevatron. My primary research interest was particle hadronization in quarkonium production in proton-antiproton collisions.

byob !

DSD !

Notes: :)

1. This whole 'notes' joke, is getting *pretty* tired . . . more content !!!

Acknowledgements:

A great, big shout-out to the friendly professional staff at CreateSpace.

They take my scrappy little pdf files and make them look like books.

All for free !

Thanks to Amazon for providing this platform, and promoting my books.

Thanks to Google Books for recognizing my books as such.

These guys (they're people now!) may be taking over the world,

but,

it seems they genuinely care about books,

and that is cool.

As for the robots taking over the world --

Do not worry.

That is, like, *the* most **stupidest** idea ever.

: /

easter egg:

a qm toe

!!!

chill

do					be

go

there is no will

wot ???

d'oh !!

there is no try

no!   lies!

just now

now gone !

surprise

gf

Toward a Metaphysics of Mass and Motion

:)

Greg Feild

October 29, 2018

boo!

About the author:

    I earned a PhD in experimental high energy physics from the Pennsylvania State University working on HERA at DESY in Hamburg, Germany studying photoproduction and deep inelastic scattering in electron-proton collisions.

    I did my postdoctoral studies with Yale University working at Fermilab on the CDF experiment at the Tevatron. My primary research interest was particle hadronization in quarkonium production in proton-antiproton collisions.

    My dogs are Jake and Seal.
They were named by Rescue Organizations. (they're people now!)

---

time the avenger:

nobody's perfect
not even the perfect stranger
but oh what a gal
she was such a perfect stranger
and you're the best in your field
in your office with your girls
and desk and leather chair
thought that time was on your side
but now it's time the avenger

---

time, time, time
hear the bells chime
over the harbor
and the city
one more vodka and lime
to help paralyze
        that tiny little

                              tick, tick, tick

                              -- Chrissie Hynde

Abstract:

In this book, we see if we can conceive of a metaphysics
based on the physical concepts of mass and motion

                          Most people only use ten percent of their brain.
                          I am now one of those people.

                                                       -- Bart Simpson
                                                        The Simpsons

                                                                Things 'no one' understands:

                                                                      quantum mechanics
                                                                           astrology
                                                                             dowsing

I am a child:

              you make the rules
              you say what's fair

                        it's lots of fun   to have you there

god gave to you   now
you give to me

I'd like to know
                                                                             what you've learned

the sky is blue
and so is the sea

what is the color
when black is burned?

                                          what is the color ?

                                                                    -- Neil Young

Metaphysic:

   1 a : METAPHYSICS
   2   : the system of principles underlying a particular study or subject
      : PHILOSOPHY

Metaphysics:

   1 a (1) : a division of philosophy that is concerned with the fundamental nature of reality
         and being and that includes ontology, cosmology, and often epistemology
     (2) : ONTOLOGY
    b : abstract philosophical studies : a study of what is outside of objective experience
   2   : METAPHYSIC 2

Mathematical:

   1   : of, relating to, or *according with* mathematics   [italics, ours]
   2 a : rigorously exact : PRECISE
     b : CERTAIN

Physics:

   1 : a science that deals with matter and energy and their interactions

Physicist:

   1   : a specialist in physics
   2 *archaic* : a person skilled in natural science

-- *Merriam Webster's Collegiate Dictionary Eleventh Edition*

rif

On mysticism:

How do we know someone is totally enlightened and has achieved inner peace ?

How do we know someone has, or s about to, unlock the secrets of the universe ?

Only on their say so.

On predeterminism:

Imagine someone who goes to the chiropractor. They were raised going to the chiropractor. Their parents took them to the doctor, the dentist, the eye doctor, and the chiropractor. They were born and bred to go to the chiropractor.

As an adult, they are free to choose to go or not to go to the chiropractor, but they are not quite so free to consider the reality of the chiropractor in their decision.

Even presented with incontrovertible evidence that the chiropractor is not a real thing; (most of) these people will not be able to believe it.

This is cognitive dissonance. (see also, celebrity physicist) See also, Indoctrination.

On information:

celebrity physicist ⇨ zero information transfer

On quantum information:

celebrity physicist ⇨ not one *iota* of information transferred

On time travel:

Recently U.S. politicians were involved in a grisly debate over who would get into a time machine and kill the most babies.

                                                    Disturbing !

The time machine is impossible.

Metaphysics: Greek, "after the physics", *meta* - after, beyond, *physikos* - pertaining to nature, *physis* - nature

The origin of the word *metaphysics* is uncertain. Aristotle did not use the term, although there is a compilation of his works called *The Metaphysics*. There is no general agreement as to how to define metaphysics. The following are some of the main definitions:

1. Metaphysics is the attempt to present a comprehensive, coherent, and consistent account (picture, view) of reality (being, the universe) as a whole. In this sense, it is used interchangeably with SYNOPTIC PHILOSOPHY and COSMOLOGY.
2. Metaphysics is the study of BEING as BEING and not "being" in the form of a particular being (thing, object, entity, activity). In this sense it is synonymous with ONTOLOGY and FIRST PHILOSOPHY.
3. Metaphysics is the study of the most general, persistent, and pervasive characteristics of the universe: existence, change, time, cause-effect relationships, space, substance, identity, uniqueness, difference, identity, unity, variety, sameness, oneness.
4. Metaphysics is the study of the ultimate reality -- reality as it is constituted in itself apart from the illusory appearances presented in our perceptions.
5. Metaphysics is the study of the underlying, self-sufficient ground (principle, reason, source, cause) of the existence of all things, the nondependent and fully self-determining being upon which all things depend for their existence.
6. Metaphysics is the study of a transcendent reality that is the cause (source) of all existence. In this sense, metaphysics becomes synonymous with THEOLOGY.
7. Metaphysics is the study of anything that is spiritual (occult, supernatural, supranatural, immaterial) and which can not be accounted for by the methods of explanation found in the natural sciences.
8. Metaphysics is the study of that which by its very nature must exist and cannot be otherwise than what it is.
9. Metaphysics is the critical examination of the underlying assumptions (presuppositions, basic beliefs) employed by our systems of knowledge in their claims about what is real.

SYNOPTIC PHILOSOPHY: The attempt to envision in an abstract way an all-inclusive world view and to see the relationships of all things with one another in accordance with basic principles of change and activity.

COSMOLOGY: 1. The study of the universe as a rational and orderly system. 2. Sometimes used synonymously with METAPHYSICS: The study of the most pervasive concepts that can be applied to the universe (such as space, time, matter, change, motion, extension, force, causality, eternity). 3. Often used to refer to that branch of science, specifically a section of astronomy, which attempts to hypothesize about the origin, structure, characteristics, and development of the physical universe on the basis of observations and scientific methodology.

-- Peter A. Angeles / *Dictionary of Philosophy*

Metaphysics (Aristotle):

1. Metaphysics is the study of Being-as-such (Being-in-itself) as distinct from the study of particular beings that exist in the universe. Biology studies the "being" of living organisms; geology studies the "being" of the earth; astronomy studies the "being" of the stars; physics studies the "being" of natural change, movement and development. But metaphysics studies the properties that all these "beings" have in common. In this sense of metaphysics the most important questions are : "What is Being", "What is Substance", "What is Reality?"
2. Metaphysics is the study of what it means to say something "is", what it means "to be." Metaphysics is the study of those properties (characteristics) a thing must have to be in change and to have an identity.
3. Metaphysics is the study of the eternal First Principles (Laws) in accordance to which all things act.
4. Metaphysics is the study of the separate realm of eternal, unchanging Being. In this sense metaphysics becomes identical with the traditional definition of THEOLOGY.
5. Metaphysics is the study of nonsensible (insensible, not sensed) substance as opposed to the sciences that deal with sensible substances. Aristotle referred to 4 and 5 as FIRST PHILOSOPHY.
6. Metaphysics is the cataloging of (**a**) the general level of realms, of things that exist which are dealt with by the sciences, and (**b**) the study of how such realms of existence relate to one another and how they provide the framework in which activity occurs and by which it is limited.
7. Metaphysics is (**a**) the study of the interrelationships of all types of knowledge, (**b**) the study of how their concepts apply (or can be intelligently applied) to what exists, and (**c**) the study of their ontological *and* logical status in providing us with truth about reality.

Metaphysics as understood in **6** and **7** deals especially with such things as the ontological and logical status of universals the relationship of universals to particulars, the status of the concepts of unity, energy, change, form, mathematical points, lines, geometric forms, etc.

-- Peter A. Angeles
Dictionary of Philosophy

We thank Professor Angeles (go lions!) for his most excellent summaries.

rip

Beneath the Sinister Universe !:

        nothing

What is a physicist ?:

    Thomas Kuhn said it.
    Heidegger said it.
    Sartre said it.
    Everybody's saying it!

    The physicist is a trained technician, apprenticing with their mentor,
    learning the ropes, the paradigm, and the tricks of the trade.

On heros:

    How did people who can't reason their way out of a wet paper bag become our leaders ?

        We aren't just talking about physicists, here.   sad

        If you'd like to be an Influencer, do someone a kindness.

On turtles:

    The *place* where the story happened was a world on the back of four elephants perched on the shell of a giant turtle. That's the advantage of space. It's big enough to hold practically *anything*, and so, eventually, it does.
    People think that it is strange to have a turtle ten thousand miles long and an elephant more that two thousand miles tall, which just shows that the human brain is ill-adapted for thinking and was probably originally designed for cooling the blood. It believes mere size is *amazing*.
    There's nothing amazing about size. Turtles are amazing, and elephants are quite astonishing. But the fact that there's a big turtle is far less amazing than the fact that there is a turtle *anywhere*.

                                                -- Terry Pratchett
                                                The Last Hero

Introduction:

In the universal model, elementary particles are mathematical points of/with angular momentum. The greater the angular momentum; the greater the mass (energy) of the (point) particle, the greater the velocity, and the greater the 'linear momentum' of the particle.

These point particles interact by emitting discrete, fundamental units of angular momentum (of varying energy), Planck's quantum of action, h, otherwise known as the photon; *or* by *exchanging* angular momentum via a 'virtual photon', of energy, $E = h/t$, where t is the time of the interaction.

In the universal model, there are only *interactions* and no *events*.
        There are only **moments** of *some* duration.

This would be our <u>metaphysic</u>, in a nutshell. As for Metaphysics … so many kinds !

Every interaction conserves angular momentum, and the sum of the change in energies of the objects involved is always zero; so in this sense, in the universal model, there is never any fundamental change, ever! This is a fun observation, but not of much practical use.

First, we note that metaphysics is not a 'meta' language in which to discuss the rules of physics, such as the prefix is used in the term 'metamathematics.'

Let's look to our list of possible definitions and choose the ones that seem most appropriate for describing our physical and mathematical investigations.

From the first list, for mere mortals, we choose **1**, **3**, **8**, and **9**, and declare the universal model to be a synoptic (a new word to me!) philosophy and a cosmology.

The world is comprised of interacting 'quantums of action'; time and space are Newtonian; the universe is static and eternal.   Clear and simple.

From 'Aristotle's' list, we choose **2**, **3**, and **7**.

The fundamental unit of matter is the neutrino, and in a sense it gives rise to the electron and the photon. These three particles are *defined* as **mass in motion** and they combine to form all matter. The impenetrability of matter is due to fast moving, closed electronic orbits.

    Seriously!   Imagine trying to stick your hand into a spinning fan blade.

                            ouch

What is mass ?:

In the universal model, any particle's charge (i.e. gravitational mass) is equivalent to its inertial mass. Previously, we have referred to the photon as "inertialess", but we now see this was a misnomer (i.e., a mistake in reasoning and the use of words!).

We have also said that there are two kinds of mass: matter and light (EM radiation).

The photon is not considered matter, but it *is* inertial mass. It takes the same amount of force to accelerate an electron or a photon (i.e, change its *frequency*) with equal kinetic energies. Inertial mass, or the resistance to acceleration, is ultimately due to the conservation of angular momentum.

Remember, in our model, photons gyre and electrons gimble; two sorts of simple harmonic oscillation with a quantum twist. The twist being that elementary particles change polarization every 2π radians as the angular momentum rotates about the direction of travel. This is the source of a particle's linear momentum, or its ability to deliver an impulse, or force, to another object, upon collision.

Just for fun, and to fill this page we reproduce our picture of photon propagation in Figure 1.

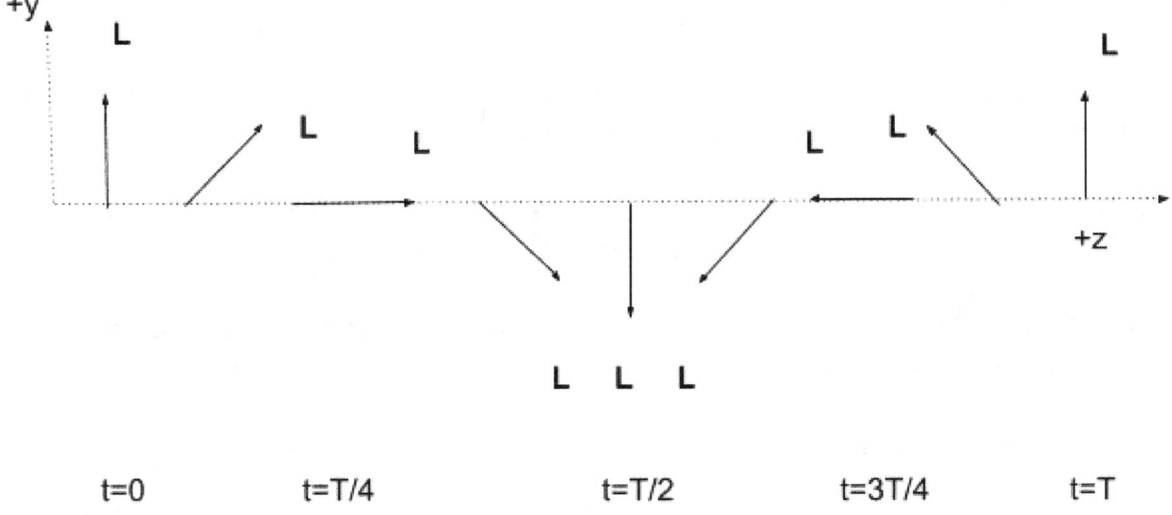

Figure 1: The photon rolls, or 'gyres', along the direction of travel projecting the spin angular momentum vector sinusoidally along the direction of propagation while maintaining a constant polarization and angular momentum; L = h. We could also say that the photon '**spirals**'. At t=0, the plane of the photon spin is *coplanar* with the z-axis.

On reason and rationality:

    According to all the great philosophers, reason and rationality are slippery and elusive concepts, so we are not going to try any definitions today.

    However, I say reason and rationality are like obscenity; I may not be able to define what a reasonable argument or concept must look like and all the logical criteria it must meet, but I know one when I see one.

    Unfortunately, everyone else feels the same way. Even wrong people!

Throwing humility aside, over our last twenty books we've seen good reasoning and bad reasoning, and the recognition and rectification of the bad reasoning after further thought. It took ten books from the time we introduced the neutrino magnetic moment until we finally got it right. This is only one example of several; e.g. the model of muon decay!

    Yes, we do think these things are right. Hence, all the books.   :)

    It's all about trying to form *clear and simple* ideas. Each clear and simple idea *must lead irrevocably* to the next.

    If not, stop. You are at the end; or at least an impasse. Wait patiently. Don't make crap up.

On belief:

    As we have seen, people will believe *anything* (e.g. broccoli cures cancer, vaccines cause autism, spacetime is curved and expanding). These beliefs are always based on some persons' alleged <u>authority</u>; be it a quack doctor or a crank physicist.

    They're just people, yall, trying to make it some way; just like you and me.

    Remember what David Hume said on miracles; and we paraphrase --

        If it sounds too good to be true, it probably is!

        There's one born every minute.

                 -- P. T. Barnum

Conservation of energy and momentum:

'Classically', and non-relativistically, the relationship between a moving particle's three *linearly independent* momenta and its energy is

$$(p_x^2/2m_0 + p_y^2/2m_0 + p_z^2/2m_0) = K = E \tag{1}$$

The allowed values for the momenta are 'infinitely' degenerate. Imagine a cosmic ray traveling through interstellar space with energy E. Each of the three momenta can take on any continuous value from zero to $\sqrt{2mE}$.

Relativistically, and 'classically', equation (1) takes a quite similar form (19,20);

$$(p_x^2/m + p_y^2/m + p_z^2/m) = K = E - mc^2 = mc^2 - m_0c^2 \tag{2}$$

$$p^2/m^2 = (1 - m_0/m) c^2 \ ; \ p^2/m^2 = v^2 = (1 - m_0/m) c^2 \tag{3}$$

we also know, $p^2/m = pc$, and $p = h/\lambda$ so

$$pc/m = (1 - m_0/m) c^2 \tag{4}$$

$$\rightarrow \ h/mc = \lambda (1 - m_0/m) \tag{5}$$

Equation (5) defines the 'relativistic Compton wavelength', λ_{rel}

$$\lambda_{rel} = (h/mc) / (1 - m_0/m) \tag{6}$$

For an accelerating particle, starting from rest

$$\Delta \lambda = \lambda_{rel} - \lambda_0 \tag{7}$$

algebra is fun

Particle representation:

The 'electromagnetic charge' and the 'mass charge' were designed to look like a '4-vector', although the pair do not form a Lorentz invariant scalar,

$$Q_{EM} = e \hbar / 2c \, \mathbf{s} \qquad (8)$$

$$Q_{MASS} = m \hbar / 2c \qquad (9)$$

although these quantities must be conserved at every 'vertex'.

We should be able to form a covariant 4-vector from the pair (**L**, I) -- something like;

$$L_x^2 + L_y^2 + L_z^2 - c^2 I^2 = 0 \qquad (10)$$

Let's check it out! In "On Rotation", we discovered the angular momentum and the moment of inertia of a particle obey the following (cut, pasted, and amended) relationships

$$I = m \lambda^2 / (2\pi)^2 \qquad (11)$$

$$T = I \omega^2 \qquad \text{[the factor of ½ is gone]} \qquad (12)$$

We define the 'radius' of a point particle to be equivalent to the wavelength. This radius travels 2π radians every wave cycle. The moment of inertia is dependent on the particle wavelength, $I = I(\lambda)$.

$$L = \sqrt{3} \hbar / 2 + I \omega \qquad (13)$$

$$E = m_0 c^2 + I \omega^2 \qquad (14)$$

So, we define the 'four-angular-momentum vector' for the photon (→ lepton) to be

$$L_{CHARGE} = (I \omega \mathbf{s}, I \omega^2) \rightarrow \text{mass} \rightarrow (L \mathbf{s}, E) \qquad (15)$$

                photon   **lepton**

where L and E are given by equations (13) and (14), and **s** is the unit spin vector. We trust the reader will want to do the math, and find that the squares of the two 4-vectors of equation (15) yield the proper and desired results. (too many Ls and Qs !)

homework!   :/

The double slit experiment:

In our model, the electron is a real, physical object.

In the sinister universe, if something can put your eye out, it is real.

As we have seen, the electron has an extension, or 'radius', that is velocity (i.e wavelength) dependent. The mean cross sectional area of an electron will then be dependent on the running of α: $\sigma_e(v) \sim \alpha(v)^2$.

In a double slit experiment, an electron with a cross section on the order of the slit separation, $\sigma(v) \sim d$, will clip both slits, and like a perchenko ball, pass through one or another.

Physically, or mathematically, it seems hard to paint a clearer than this.
Let's say each electron has a 50-50 chance of passing through one slit or the other slit.

Each slit then acts as a point source of electrons, with the usual spread in momentum due to the uncertainty principle.

Each individual electron arrives either 'in phase or out of phase', for darkening a photographic plate, just as it does in our explanation of the single slit experiment (11).

Note: We've had several less good ideas on the way to this explanation of double slit interference. I think this is the *one*.

keep trying

keep *thinking* !   :)

The standard model:

It's dead, Jim.

The Dirac equation:

This is our 21st book on grand unification, and we are *still* hunting down hidden factors of hbar and c. It's easy to blame everyone else, so we will! We found the last (?) suppressed factor of c in "Quantum Field Theory" by F. Mandl and G. Shaw; although no book is entirely internally consistent on when they choose to show c or set c = 1.

So, the universal Dirac equation is *now* (corrected from "On Quantum Mechanics")

$$H \psi = c \alpha \cdot p \, \psi \tag{16}$$

$$i \hbar \, \partial \psi / \partial t = -i c \hbar \, \alpha \cdot \nabla \psi \tag{17}$$

and is satisfied by, or solved with, the universal wave function

$$\psi = \exp(i(p \cdot x - (m-m_0)c^2 t)/\hbar) \tag{18}$$

$$mc^2 \psi = c \, \alpha \cdot p \, \psi + m_e c^2 \psi \tag{19}$$

The free particle, spin up, solution of the Dirac equation for an electron (positive helicity) is

$$\psi = \begin{vmatrix} 1 \\ 0 \\ \sigma \cdot p/mc \begin{vmatrix} 1 \\ 0 \end{vmatrix} \end{vmatrix} \exp(i(p \cdot x - (m-m_0)c^2 t)/\hbar) \tag{20}$$

$$\sigma \cdot p/mc = \begin{vmatrix} p_z & p_x - ip_y \\ p_x + ip_y & -p_z \end{vmatrix} \tag{21}$$

→ the determinant = $-(p_z^2 + p_y^2 + p_x^2) \rightarrow -(p_z^2 + p_y^2 + p_x^2)/mc = -1$ (22)

→ the square = $(p_z^2 + p_y^2 + p_x^2) \mathbf{1}$; $\mathbf{1}$ = 2x2 unit matrix (23)

The matrix $\sigma \cdot p/mc$ *is the factor* that turns our ordinary column vector *into a spinor*, which is technically classified as an Euclidean tensor. So the universal model involves scalars, vectors, and tensors: all Euclidean.

A spinor is a vector with zero 'length' which means that two of the three components must be imaginary. In our example, only p_z is real; the direction of propagation!

The form of equation (21) turns up again and again; i.e. rotations, angular momentum ... however, equation (21) is not the result of a rotation or 'cross product', as in the other examples.

let's all stew on this thought !

The Dirac equation with the universal interaction term added is

$$i\hbar \,\partial\psi/\partial t = -ic\hbar\, \alpha \cdot \nabla \psi + mc^2 A^\mu \tag{24}$$

where A^μ represents the wave function of the virtual photon mediating the interaction with a second particle. <u>It does not represent a potential field.</u> For a single particle interacting with several other particles, A^μ would be represented by a Fourier sum or integral over all the relevant virtual photon frequencies.

no more fields

conclusion: a poem

everything is normal
everything is real

all

the *more* reason

for crazy feels

On gravity:

Newton's universal law of gravitation for the force between two bodies or objects, normalized by the total relativistic energy of the two bodies, and expressed in the center of mass of the two body system, is

$$F/E_{TOT} = K*(c/R)^2 \mu - K*(\mu v^2/R^2) - K*(l^2/\mu R^3) \quad (25)$$
$$K = G/c^2 \quad (26)$$

The second term on the right hand side of equation (25) is the *coriolis* force; our answer to spacetime disturbances. The inertial coordinate system is fixed with respect to the background of empty space, so the coriolis effect is *solely* due to the relativistic velocity of the objects, and not due to the choice of any particular reference frame.

Were people ever really satisfied with the explanation for ' why the coriolis force seems to act **in the wrong direction?** '

Yo -- there is still magic, mystery, and things to discover in the classical realm !

Remember the potential energy; a great and glorious concept until people made it real. This 'event' seems to have happened organically; people talked about potentials all the time, they were very useful in rigorous calculations and for hand waving explanations, and they could be measured !   Like the temperature of a system of bodies, $E = 3/2\, nkT$, which can **also** be measured, potential energy is mass in motion.

Naturally, the reality of this *magical potential energy* was assumed in the foundation of quantum mechanics; thus dooming it to **mysticism**, and failure, from the very start.

The angular momentum of a two body planetary system is constant. In our model, this means the bodies *are not accelerating*. Bodies only radiate when they experience a change in angular momentum. Hence, the motion of the bodies in our solar system, for example, will generate ('multipole') magnetic fields, but **no** gravitational waves.

As in the case of classical, Newtonian gravity, we will assume that after integrating, we can represent the force as one virtual photon acting between the centers of the two bodies. The energy and momentum transferred by the photon will then depend only on the change in radial separation; dr/dt.

For circular, two body motion ($m_1 = m_2$), the virtual photon has energy, $E = 0$, and *only* carries units of h; angular momentum!

$$dr/dt \Leftrightarrow \Delta E \;;\; d\theta/dt \Leftrightarrow \Delta L \;;\; \Delta L \equiv n h$$

reality is fun !

Resources:

Quantum Field Theory
Claude Itzykson, Jean-Bernard Zuber

Atomic and Quantum Physics
H. Haken, H.C. Wolf

Modern Elementary Particle Physics
Gordon Kane

Classical Dynamics of Particles and Systems
Jerry B. Marion

Foundations of Electromagnetic Theory
John R. Reitz, Frederick J. Milford, Robert W. Christy

Quantum Physics
Rolf G. Winter

Gauge Theories in Particle Physics
I. J. R. Aitchison and A. J. G. Hey

Quarks and Leptons: An Introductory Course in Modern Particle Physics
Francis Halzen, Alan D. Martin

Quantum Field Theory
F. Mandl, G. Shaw

Theoretical Mechanics of Particles and Continua
Alexander L. Fetter, John Dirk Walecka

The Theory of Spinors
Elie Cartan

Elementary Modern Physics
Richard T. Weidner, Robert L. Sells

Quantum Mechanics
Claude Cohen-Tannoudji, Bernard Diu, Franck Laloe

Books by Greg Feild:

1. "A quantum mechanical theory of gravitational interactions"
   CreateSpace Independent Publishing, 8/29/2016

2. "Observations on the quantum mechanical nature of gravity"
   CreateSpace Independent Publishing, 10/8/2016

3. "On gravitation and electric charge"
   CreateSpace Independent Publishing, 10/29/2016

4. "On spin, mass, and charge"
   CreateSpace Independent Publishing, 11/29/2016

5. "On angular momentum, acceleration, and absolute motion"
   CreateSpace Independent Publishing, 1/1/2017

6. "The Sinister Universe"
   CreateSpace Independent Publishing, 3/1/2017

7. "On Parity and Isospin"
   CreateSpace Independent Publishing, 4/11/2017

8. "Reflections on the Sinister Universe"
   CreateSpace Independent Publishing, 5/12/2017

9. "On Current Physics"
   CreateSpace Independent Publishing, 6/11/2017

10. "A Critical Examination of Classical and Quantum Mechanical Waves"
    CreateSpace Independent Publishing, 6/18/2017

11. "On wave particle duality and the quantum of action"
    CreateSpace Independent Publishing, 7/6/2017

12. "On matter, mass, and motion"
    CreateSpace Independent Publishing, 9/14/2017

13. "On action and reaction"
    CreateSpace Independent Publishing, 9/24/2017

14. "A quantum mechanical theory of everything"

CreateSpace Independent Publishing, 11/5/2017
15. "On Interaction"
    CreateSpace Independent Publishing, 4/21/2018

16. "On Rotation"
    CreateSpace Independent Publishing 8/19/2018

17. "Revenge of the Sinister Universe: The Reality of Everything'
    CreateSpace Independent Publishing, 9/4/2018

18. "On Math, Physics, and Metaphysics"
    CreateSpace Independent Publishing, 10/1/2018

19. "On Quantum Mechanics"
    CreateSpace Independent Publishing, 10/15/2018

20. "On Epistemology and Ontology"
    CreateSpace Independent Publishing, 10/21/2018

21. "Toward a Metaphysics of Mass and Motion"
    CreateSpace Publishing, 10/29/2018

Compilations:

A. "The Universal Model of Our Sinister Universe: The First Ten Books"
   CreateSpace Independent Publishing, 7/2/2017

B. "The Canons of the Sinister Universe:
   The Last Four Books on the Universal Model of Our World"
   CreateSpace Independent Publishing, 11/5/2017

C. "The Return of the Sinister Universe: The Immaculate Collection"
   CreateSpace Independent Publishing, 9/4/2018

D. "The Battle for the Sinister Universe: The Heuristics"
   CreateSpace Independent Publishing, 10/29/2018

Contest: Take 21 and D; make your best Set Theory Joke (some are hilarious!)
         and send it to your favorite Celebrity Mathematician.

         If you can't find one of those, your idea here !

resistance is futile

Things that are not like computers:

    the universe

    brains

    things that are not computers

These troubled times:

In these troubled times, we don't need people who don't know what they're doing, telling other people what to do.

The post truth era:

Can physics lead us forward from our post truth apocalypse; *its* own creation, and *our* Frankenstein's monster ?

Only with contrition and humility.

nice knowing everyone

you will be assimilated

On grammar and the neutral personal pronoun:

    We've been using the third party plural to refer to unknown individuals and people in general since Mother taught us to do so in the sixties.

    An idyllic age. We were five -- and already using the royal we.

The metaphysics of repose:

    It takes energy to do stuff:

        kinetic energy
        chemical energy
        biological energy --- where are our crank theories on the ATP cycle ?

                it is mind boggling !

                and beautiful

                and it keeps us alive

:/

the windmills of your mind: #thereisonlynow

      round

      like a circle in a spiral
      like a wheel within a wheel

      never ending or beginning
      on an ever spinning reel

      like a snowball down a mountain
      or a carnival balloon

      like a carousel that's turning
      running rings around the moon

      like a clock whose hands
                               are sweeping
      past the minutes of its face

      and the world is like an apple
                                                   swirling silently in space

            like the circles that you find

                                                      in the windmills of your mind

[heavy]

                                                -- Alan Bergman, Marilyn Bergman; Michel Legrand

www.ingramcontent.com/pod-product-compliance
Lightning Source LLC
Chambersburg PA
CBHW062332220526
45469CB00008B/2682